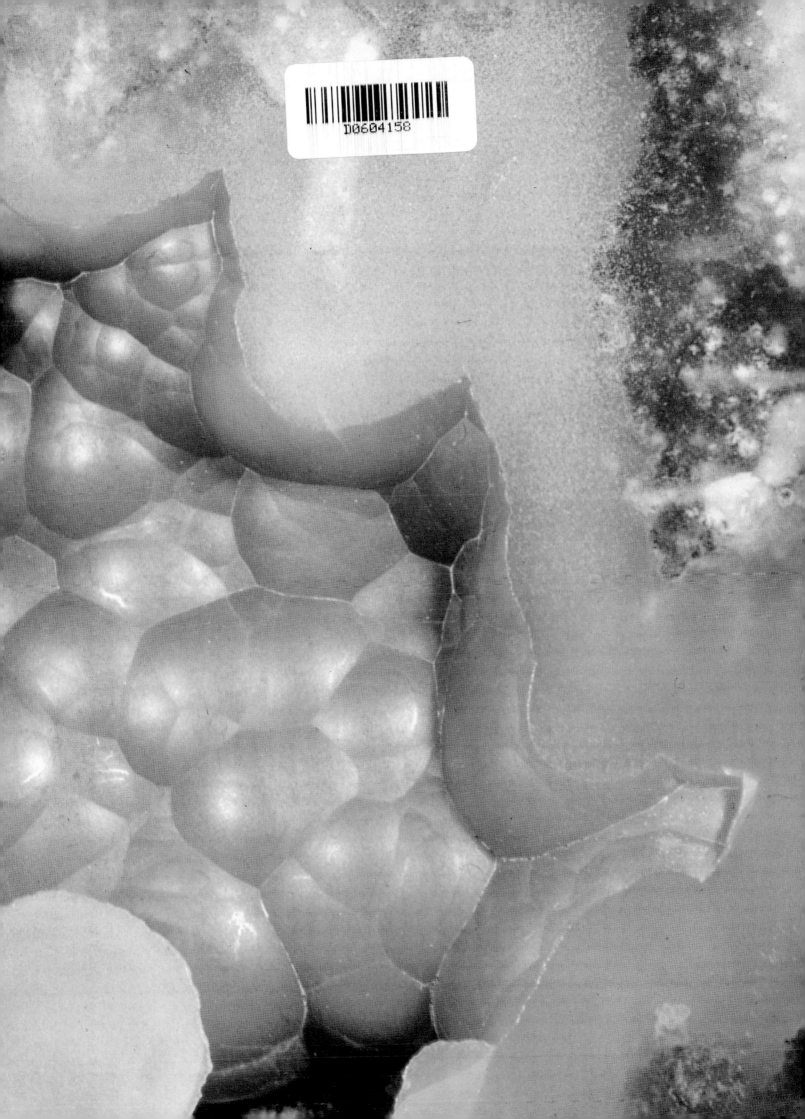

THE
FOSSIL
WORLD

THE FOSSIL WORLD

Richard Moody

CHARTWELL BOOKS INC.

Preface

A book of this nature is dependent to a great extent on the work carried out by dedicated researchers throughout the world. The text is but a summary of the vast amount of information that is published each year. It is, however, an attempt to combine this information with the observations and ideas of the author, who is responsible for the manner of the data and interpretations presented in this book.

The author wishes to thank Mr G. Raine, Mr D. Wright, Miss S. Cornish and Mrs F. Mouton for their help in the preparation and presentation of this book; and the British Museum (Natural History) and Kingston Polytechnic for allowing their specimens to be photographed.

Line drawings by Linda Parry.

Published by
The Hamlyn Publishing Group Limited
London·New York·Sydney·Toronto
Astronaut House, Feltham, Middlesex, England

Published in the United States by
Chartwell Books Inc.,
A Division of Book Sales Inc.,
110 Enterprise Avenue
Secaucus
New Jersey 07094

Copyright ©
The Hamlyn Publishing Group Limited 1977
ISBN 0 600 33609 3
ISBN 0-89009-116-1 (Chartwell edition)
Library of Congress Catalog Card No. 77-71223

Phototypeset by Filmtype Services Limited,
Scarborough, England
Printed in Hong Kong

The sizes given in the captions refer to the particular specimen and are approximate.

Contents

Introduction

Between the writings of Xenophanes (5th century BC) and Herodotus (4th century BC) and the development of palaeontology as a modern science, frequent references were made to 'outgrowths from the earth's surface' and petrifactions of plant and animal remains. The word *fossilia* in Latin means anything dug up or extracted from the earth and terms such as *fossilia nativa* and *fossilia petrificata* were used to describe minerals and organic remains.

In modern times the word 'fossil' has come to mean the remains of plants and animals preserved in sedimentary rocks deposited during the passing of geological time. The fossil itself may be simply a cast or mould, a fragment of the original skeleton, the whole skeleton or, very rarely, the complete organism. It may, in turn, be an impression of a leaf, the footprint of a dinosaur or the traces of animals involved in feeding, burrowing for protection or boring into a hard substrate. The study of fossils or ancient life is called palaeontology, and is a science founded on observation and description. It is a progressive science, moving more in recent times towards the analysis of faunas and floras and the interpretation of past environments. The science has to some extent been subdivided into palaeobotany, the study of fossil plants, palaeozoology, the description and analysis of fossil animals, and palaeoecology, the interpretation of ancient organisms and their environments. Further specializations may involve palaeontologists in the dating of rocks, the reconstruction and function of skeletons, and the distribution of fossils in space and time.

The palaeontologist will be involved in the description of individual fossils, their associations with other organisms entombed within the enclosing sediment and the sediment itself. Sedimentary rocks are formed by the action of the various mechanical and chemical processes of nature on existing rocks. The products of weathering are deposited in various environments such as deltas, lakes or seas. The size of the particles which make up the sediment and the structures created during deposition provide information on the palaeoenvironment. The preservation of fossils will depend largely on the mechanical and chemical processes active during and after the deposition, and the chances of fossilization of any organism that dies and settles on the substrate are greatly increased by rapid burial. One only has to look at a modern-day beach to see the destructive effect that waves have on shells. Soft parts of animals and plants are rarely preserved in the fossil record, attack by bacteria, scavengers or the action of the various elements causing breakdown and decay. Naturally, exceptions do occur, particularly when conditions exist which prevent the action of bacteria and scavengers. The entombment of insects in the fossil resin amber, or of mammoths and rhinoceroses in frozen sediments are examples. Tar lakes, peat-bogs and brine-saturated sediments have all yielded excellent fossils. Mostly these are young in the geological sense, but specimens such as the rhinoceros of Starunia or the 'Bog People' of Aarhus provide an insight into fossilization.

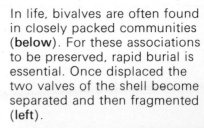

In life, bivalves are often found in closely packed communities (**below**). For these associations to be preserved, rapid burial is essential. Once displaced the two valves of the shell become separated and then fragmented (**left**).

Where the original shell material has been dissolved, casts and moulds may remain to provide evidence of the form of the animal. Internal moulds of gastropods and bivalves from Portland stone, Upper Jurassic, England. 11·5 cm (4·5 ins) long

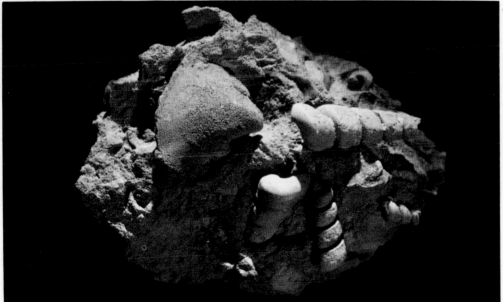

More common are the impressions of delicate organisms preserved in the finer-grained sediments. The carbon-rich deposits of the Messel region of Germany and the bituminous shales of Menat in the Auvergne region of France have yielded many beautiful fossils, impressions of leaves, beetles and fish being common to both outcrops. These sediments were laid down over forty million years ago, but soft-part impressions have been found in sediments of even greater antiquity. The lithographic limestone of Solnhofen (Jurassic) in Germany and the Burgess shales (Cambrian) of Canada yield older, perhaps even more spectacular, collections.

Of the organisms that live on the sea-floor at the present time, the vast majority have no hard skeletal parts. Therefore, the chance of their being preserved in the fossil record is slight. Even those with hard parts have poor chances of being preserved and only a minute percentage will survive the

Organic materials, such as leaves, are often preserved as carbon replicas. This picture shows a Recent specimen which could be preserved in this way.

Gastropod shell replacement by opaline silica. 1·7 cm (0·6 ins) long.

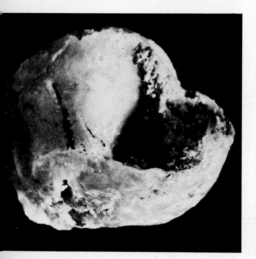

agents of destruction. The same element of chance has existed throughout geological time, and when coupled with the possible chemical changes that can take place after burial and consolidation, erosion and the re-working of sediments, the percentage of organisms preserved is reduced even further. It is, however, a limited exercise to think of the organisms that have vanished without trace; the study of body fossils and the tracks and trails of past life offer a more absorbing pastime.

The hard parts of organisms that are preserved in sediments may be composed of chitin, keratin, calcium carbonate, calcium phosphate, silica and wood. Sometimes they are found unaltered, even after 560 million years, but mostly they have been changed in some way. Minerals in solution may add to or replace the original constituents of the hard parts. The process of *petrifaction* occurs when secondary minerals impregnate fossil remains, increasing the weight and hardness of the specimen and, as the word suggests, turning the hard parts to stone. *Recrystallization* of fossil hard parts seldom affects its form but the chemical changes involved may completely alter the microstructure of the skeletal layers. In some sediments the percolation of pore waters may be sufficient to dissolve out the original minerals, and, depending on the nature of the specimen, leave moulds and casts within the rocks. Sometimes minerals within these waters will replace the original remains. *Replacement* minerals often result in fossils of enhanced beauty, particularly when silica replaces wood, or iron pyrites the shells of molluscs. In the cases of plants or chitinous skeletons the most frequent process of fossilization is *carbonization*. This process, linked with the activity of anaerobic bacteria, results in a reduction of the original chemical constituents and a fossil rich in carbon and usually very dark in colour.

Brachiopod illustrating replacement by pyrite. 4·8 cm (1·9 ins) wide.

Petrified wood from Arizona.

Coprolite. 28·6 cm (11·25 ins).

Apart from the body fossils preserved by the various processes described above, other fossils such as footprints, tracks, trails, burrows and borings provide further evidence of ancient life. These, together with faecal pellets and coprolites (the droppings of animals), are known collectively as trace fossils. They are best preserved in alternating beds of sandstone and shale and, unlike body fossils, benefit from the various processes associated with deposition and compaction of sedimentary rock formation. Trace fossils present different problems to the palaeontologist for, whilst they are invaluable as indicators of environment, they are difficult to classify in the biological sense as many organisms having the same feeding habits will leave behind similar traces.

In both cases, however, it is possible to compare some of the indicators of past life with those of the present. This involves the principle of *uniformitarianism* which can be summed up in the phrase 'the present is the key to the past'. Some genera have an incredible geological history, ranging from the Ordovician period to the present day, a span of some 500 million years, almost without change. *Lingula*, a brachiopod or lampshell, is an example of such a genus. The persistence of such forms and the continuity of similar traces throughout time enable worthwhile comparisons to be made. However, fossils with limited records and structures unknown amongst Recent organisms often offer the most exciting problems to the palaeontologist. These problems of classification, dating and perhaps reconstruction are all important, but it is difficult to understand them unless one is aware of the scales involved in geological time.

In geology, time is measured in millions of years, with the estimated age of the Earth being some 4 600 million years. Most of this time, approximately seven-eighths, is represented by the Precambrian, which was, before radioactive dating, subdivided on the degree of metamorphism of the existing rocks.

Rocks subjected to the stresses and temperatures generated within the crustal layer of the Earth, or the heat at the boundary of a magma may be deformed and recrystallized to become metamorphic rocks. The degree of metamorphism varies with the range of pressures and temperatures that prevail at a given time and, because of this, some rocks of great antiquity may be less altered than others of a younger period, which means that the dating of rocks by this method may be subject to error.

Dating using the rate of decay of radioactive isotopes trapped in the rocks has provided the geologist with the means of assigning absolute ages and has proved that dating through the intensity of metamorphism is unsatisfactory. Radioactive dating has indicated major mountain-building episodes at 2 500, 1 000 and 700 million years ago. The first date divides the Precambrian into the Archean and Proterozoic Eras, and the last two divide the Proterozoic into three major periods.

Geological time-scale and outline of the fossil record.

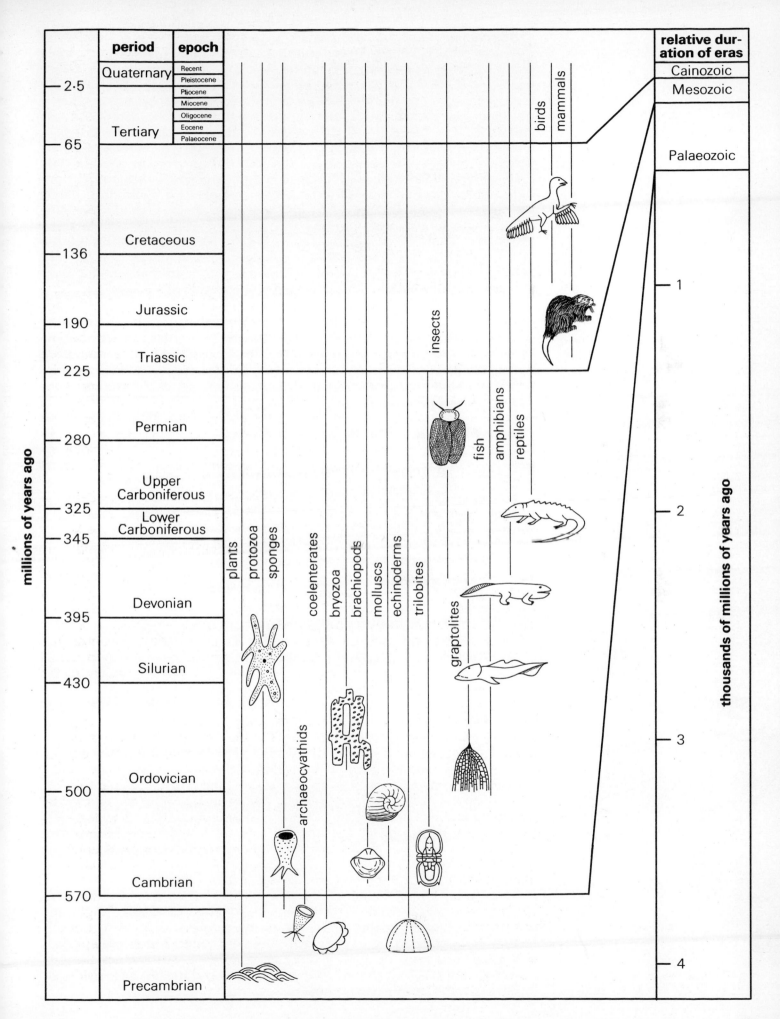

Radioactive dating has also enabled geologists to correlate the history of the Earth with that of the Moon, and through the study of meteorites, other planets. The ages of the majority of meteorites so far measured are between 4 500 and 4 700 million years. Of the various radioactive isotopes, uranium-lead, potassium-argon and rubidium-strontium are of considerable value in the dating of ancient rocks; thorium-protoactinium in the dating of rocks and fossils less than 250 000 years old and carbon for specimens less than 50 000 years old. The use of 'nuclear clocks', as they are sometimes called, is now essential to the geologist, but this method can only be employed on rocks with radioactive elements trapped in them and various other conditions must also be satisfied. Therefore the classic comparative methods of dating are still used throughout research and industry. These methods, including the changes in sedimentation and the occurrence of fossils in particular strata offer a relative time-scale. The essential clues to subdivision are the occurrence of important worldwide geological phenomena. Comparisons between the thicknesses of sediments laid down since the start of the Cambrian period and published radioactive dates show good agreement.

The summation of these various methods can be seen in the stratigraphic column, the data presented being a far cry from earlier attempts. Archbishop Ussher, for example, is accredited in 1658 with setting the creation of the Earth at 4004 BC, an age borne of biblical chronology. Another attempt at dating, based on the amount of salt in the oceans of the world compared with the amount of salt added each year, gave an age of 100 million years.

List of phyla

Kingdom	Phylum	Common Name
Protista	Cyanophyta	Blue-green algae
	Protozoa	Foraminiferids, radiolarians
Animalia	Porifera	Sponges
	Archaeocyatha	Pleosponges
	Coelenterata	Jellyfish, corals
	Bryozoa	Moss animals
	Brachiopoda	Brachiopods or lamp shells
	Mollusca	Molluscs — bivalves, snails, cephalopods, chitons, tusk shells
	Annelida	Worms
	Arthropoda	Trilobites, crabs, lobsters, insects, spiders
	Echinodermata	Sea lilies, star fish, sea urchins, blastoids, cystoids, sea cucumbers
	Hemichordata	Graptolites
	Chordata	Jawed and jawless fish, amphibians, reptiles, birds, mammals
	Division	
Plantae	Thallophyta	Bacteria, algae, fungi, lichens
	Bryophyta	Mosses, liverworts
	Pteridophyta	Spore bearing, primitive, vascular plants — horsetails, lycopsids, ferns
	Spermophyta	Seed bearing plants. Gymnosperms — seed ferns, cycads, gingkoes, conifers Angiosperms (flowering plants) — grasses, palms, trees, shrubs

Collecting and preparing fossils

In the early part of the 19th century the founding fathers of palaeontology described hundreds of new genera and species. Often, these descriptions were extremely accurate and beautifully illustrated on hand-carved blocks. Sometimes specimens were delicately prepared with hammer and fine chisel, but mainly the fine details of internal structures or of skeletons embedded in sediment went by unnoticed. Today, palaeontologists have at their disposal a vast store of knowledge that has been built up since those early days, and a host of modern mechanical and chemical preparatory methods. They are able to fully develop known material, and through a broad knowledge of the zoological and botanical sciences add new facts to existing descriptions. An article published now should not only present an amended diagnosis of the specimen, it should also relate morphology to environment and mode of life. The object of this chapter is to explain some of the techniques used in the preparation of fossils and to follow the methods employed in the identification and naming of specimens. It also presents information on the techniques and problems of fossil collecting, which, after all, is a major attraction of palaeontology. The essential urge of any palaeontologist is to find new forms and ultimately a missing link in the chain of evolution.

One of the first questions to be asked of many a palaeontologist who has made interesting new discoveries is 'Where do you look?'. The answer should reveal the amount of preparation, planning and care that goes into the work of finding fossils. The odd discovery by a casual visitor with a coal hammer is mainly a thing of the past.

Today, the professional palaeontologist or dedicated amateur will spend many long hours studying reports, maps and field data rather than haphazardly digging for fossils. Little is gained from rushing to the nearest exposure and hacking away. Most workers are specialists in a given field, either a particular group of rocks or a given group of plants or animals. Research and study will reveal where the outcrops of interest occur and what equipment and manpower will be needed to collect effectively.

The study and collection of fossils from a local claypit, quarry or coastal exposure demand the same degree of planning and care as collecting in the wilds of Mongolia, or the sands of the Sahara. The only difference between the two 'expeditions' should be in the amount of support equipment required for the one planned to take place in the remoter regions of our planet.

A good field worker will not set off without a geological hammer, chisels, brushes, pocket knife, hand lens and notebook. A selection of storage boxes, wrapping paper, and other protective items should also be packed for delicate or special fossils. Plaster of Paris and sacking, plaster bandages or polyurethane foam will be useful to protect and support specimens in transit.

Collecting is easiest where fossils have been weathered out and rest on the surface of the outcrop. Expertise acquired through experience will govern the lifting and packaging of such material, for the preservation of specimens will

Palaeontologist at the start of a dig.

vary according to the sediment in which they were entombed. Limestone or sandstone fossils are usually quite durable and just require suitable wrapping, but others from shales or mudstones may need strengthening and delicate handling. Before lifting, specimens should be photographed or sketched in situ. Where material is being collected for palaeoecological research the attitude and relationship of the specimen to those adjacent to it should also be recorded. Each fossil should be numbered and a record made of its geographical location and stratigraphic position. The latter refers to the precise position of the specimen within the outcrop, and is important because accurate correlation from outcrop to outcrop will depend on it.

In sandy sediments or in outcrops where 'picking' is easy, bulk sampling should be undertaken, for this will provide information on growth, variation and community structure. Selective sampling in such areas will lead to biased collections, for large, coloured or complete fossils will always find their way into a collecting box more easily than small or slightly damaged ones. Time and intensive collection is the only way to obtain a true selection of fossils, and bulk or block sampling the way to obtain the smallest but by no means least important fossils. Vertebrate palaeontologists, for example, use wet and dry sieves to concentrate and collect from bulk samples of certain outcrops.

Fossils embedded in hardened sediments present a different set of problems, for these have to be removed with controlled physical effort, using a sledgehammer and chisels. In the case of limestones it is often useful to exploit existing joint patterns. It is better to lift a larger than necessary block than to damage the specimen to be collected. Sandstones, phosphates and hardened shales should be chipped away from the sides and undersurface, care being taken to protect the specimen from damage and the block from fragmenting. Plastering or the application of a polyurethane jacket is essential in certain cases. Specimens should never be prepared in the field, for the tools at hand are usually unsuitable for delicate, patient work.

Fossils, collected and packed well, offer new challenges when studied in the laboratory. Mechanical and chemical methods of preparation can be employed to reveal morphological details unseen during field work. Naturally, the methods chosen must be suited for the sediments and fossils to be prepared. Sometimes specimens may be extracted from clays or sands after only a few hours of soaking in water. Others may need weeks of careful extraction by drills, air abrasives or air-dent tools. Invariably these fossils are quite hardy and unlikely to fracture or chip. Fine needles are often used to remove shaley material from fossils.

In some ways mechanical methods are little changed from those used by nineteenth-century preparators. Acid or chemical techniques, however, are a development of recent importance and ones which have proved to be invaluable in the extraction of fossils from some harder rocks. They are specialist techniques and should never be used by the inexperienced. Care must be

Below
The discovery of a fossil is followed by a period of careful excavation and preparation.

Below right
Prior to removal the fossil is protected by the application of a plaster bandage.

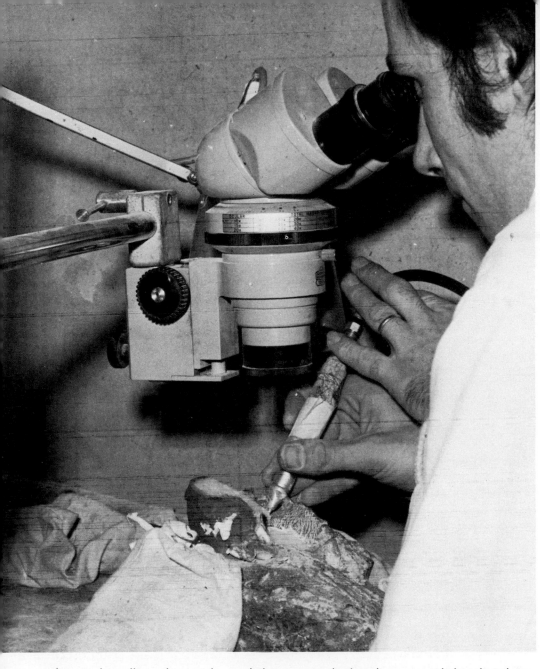

On return to the laboratory the specimen is prepared by skilled technical staff who use both mechanical and chemical aids.

taken in handling the acids, and the removal of sediment and the dissolving of minerals leave the fossil in a delicate condition. Fragmentation and collapse are the rewards for incompetence in this case.

Siliceous, chitinous or calcium phosphatic skeletons embedded in limestones can be extracted by the use of dilute hydrochloric or acetic acids. The former is best used during the preparation of invertebrate fossils, the skeletons of which have been impregnated by silica or which have chitinophosphatic shells. The specimens should be immersed in a solution of less than 10 per cent concentration. Vertebrate fossils with skeletons of calcium phosphate are best prepared in acetic acid, the concentration of which will depend on the specimen to be prepared, and the secondary minerals which impregnate the bone.

Specimens to be prepared by other mechanical or chemical methods need to be protected at all times against damage. This can be achieved by the application of a thin coating of a plastic-based glue.

Immersion in acid should be followed by washing, first in running water and secondly in de-ionized water to remove salts formed by the action of the acid on the sediments. The specimen is then dried thoroughly and newly exposed areas protected by an application of thinned-down glue. This procedure is repeated until the fossil is completely exposed. Finally, the specimen is washed for several days to remove all traces of salt and then cleaned and coated. Fossils prepared in this manner are extremely beautiful, with organisms of great antiquity resembling those of Recent times.

Apart from acetic acid, experimentation has shown that other acids are useful to the vertebrate palaeontologist; thioglycolic acid, for example, being extremely valuable in the preparation of specimens coated by iron oxides.

When collected specimens may prove to be of great scientific value, casts should be made before preparation begins. These will record the nature of the original and act as an insurance in case anything goes wrong during preparation. In the past casts and moulds were made of plaster of Paris. Now silicone rubber is used for moulds and fibreglass and resin for the casts and replicas. The end products are great improvements on past attempts, the replicas and originals differing only in weight and colour.

After collection and preparation the role of the palaeontologist changes from preparator to that of detective and curator. Fossils without names and field data are soon rendered useless, often ending up in piles in remote corners. Individual specimens should be stored carefully in boxes or drawer units and be numbered according to field area or in order of collection, and the number used as the key to a card index system. Each card should contain the field locality and systematic position of each specimen. Actual identification of individual organisms is obtained by reference to relevant texts, including field guides, published works on specific groups, or by comparison with existing collections.

Few shortcuts can be made in the establishment of a worthwhile collection. Even fewer can be employed in the complete identification and classification of individual specimens. Accurate identifications and descriptions are important, not only to classification but also to the construction of a family tree or evolutionary lineage.

Identification depends on accurate description and the selection of important features for comparison. If a new find is morphologically similar to a previously named specimen it should be possible to identify it to species level, the finder working gradually from broad or general features down to those of rather minute detail. The whole process is like opening a large box and finding smaller and smaller boxes within.

When the specimen discovered lacks previous description and can be proved to be new to the fossil record, the finder, or a specialist entrusted with the task, can classify it to the level which comparison with existing records will allow. The *specific* or *trivial* name is given to the individual specimen described and any others of close morphologic similarity. In the zoological sense a *species* is a group of organisms which can interbreed and produce fertile offspring. Several related species form a genus and a *generic* name always precedes the trivial name in any biological text. A code exists related to the naming of organisms, both living and extinct, and therefore great care is usually taken in the selection of names.

As with species, genera are grouped together to form *families*, families into *orders*, orders into *classes*, classes into *phyla* and the last into *kingdoms*. Each of these higher taxonomic groupings indicates a greater diversity of the organisms within their ranks. Three kingdoms exist, one-celled organisms (Protista), plants (Plantae) and animals (Animalia).

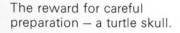
The reward for careful preparation — a turtle skull.

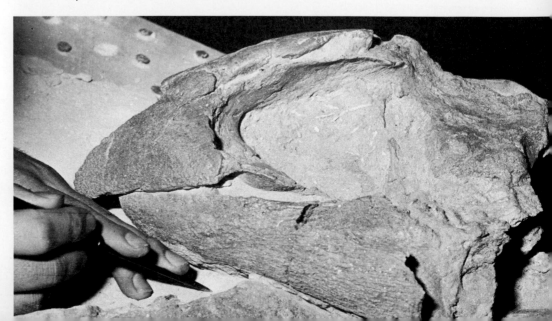

20

Precambrian life

Although seven-eighths of the Earth's history took place during the Precambrian, very little is known of the animals and plants of that eon. Research in silicified sediments has revealed that living systems existed over 3 000 million years ago and that for a vast period of time these systems remained remarkably conservative in the evolutionary sense. Around 1 300 million years ago, organisms with nuclei evolved and a series of 'advances' occurred that led to the emergence of more highly organized organisms at the end of Precambrian times.

The first organisms, recorded from silicified sediments and shales in eastern Transvaal, South Africa, are simple, rounded structures. They resemble unicellular blue-green algae. The biological simplicity of these organisms suggests that they are close to the beginnings of life and to the fusion of proteins and nucleic acids in the 'primordial soup'.

Further indications of early Precambrian life occur in the Bulawayo area of Rhodesia. These sediments are calcareous and contain laminated, mound-like structures called stromatolites, which, in this case, are the products of filamentous blue-green algae. Stromatolites are known throughout the late Archaean and Proterozoic eras, and their presence suggests that the process of photosynthesis began at an early age in the development of life. The presence of the green pigment, chlorophyll, enables plants to manufacture foods such as starch through the synthesis of water and carbon dioxide, using sunlight as energy. In carrying out photosynthesis, the earliest of plants probably changed the balance between oxygen and carbon dioxide in the atmosphere, and provided the base for future food chains and webs.

The geographic distribution of stromatolitic reefs and unicellular microscopic fossils spread during the Proterozoic and 'reefs' have been described from various sedimentary rocks in Canada, America and Australia. During the middle and late Proterozoic a new cell type evolved which possessed a true nucleus and which, after an initial period of asexual reproduction, undertook sexual combination. This event probably took place over 1 000 million years ago and sparked off a burst in the evolutionary rates amongst late Precambrian communities. It led to the evolution of higher multicelled plants and animals around 750 million years ago. From that time on diversity increased and animals and plants were to evolve which would live on the bottoms of shallow, shelf sea environments. These organisms were soft-bodied and although the record is generally poor, some faunas such as those of the Ediacara region of South Australia and the Conception Group of eastern Newfoundland, are exceptionally well preserved.

The Ediacara fauna has been dated at 600 million years old and it indicates a complex community structure. Echinoderm, jellyfish, sea-pen and worm casts are found on bedding planes within the sandstones. Over 900 specimens have been collected and the presence of ripple-marks and sun-cracks indicate a shallow-water fauna. The worms, echinoderms and sea-pens were

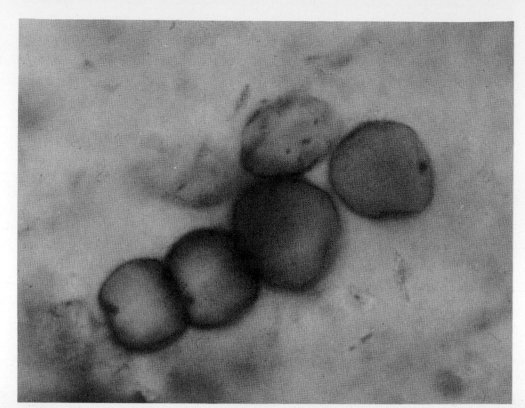

Unicellular algae from the Precambrian of Australia.
×2 500

Stromatolitic algal colonies.

bottom dwellers, whilst the jellyfish, presumably planktonic, floated in from deeper waters.

The fossils from Newfoundland are somewhat similar to those of Australia, but certain forms bear a closer resemblance to spindle and leaf-like organisms discovered in the Charnwood Forest near Leicester in England. This close similarity supports other lines of evidence indicating that eastern Newfoundland and England were joined together in these early times.

The variety of species and the number of phyla present in the Precambrian are limited, only a few phyla being represented in the Ediacara fauna. This contrasts greatly with early Cambrian communities in which some nine phyla are represented by nearly 1 000 species, many of which have hard parts. The diversity of life and the appearance of mineralized skeletons are spectacular developments in the course of biological evolution during the very late Proterozoic and earliest Cambrian times.

An explanation of the increase in diversity is probably linked with the gaps in the fossil record; important periods of evolution have been literally erased by phases of mountain building and metamorphism. The time gap between the last of the Precambrian soft-bodied faunas and the communities of the Cambrian may be of the order of fifty million years, during which time several new phyla could have evolved and organisms become more cosmopolitan, selective breeding and adaptation leading to a host of new varieties. Both animals and plants may have evolved which could occupy a number of new ecological niches.

The evolution of hard parts on the other hand, could be associated with a number of problems facing early, highly organized organisms. Some researchers believe that the ozone content of the atmosphere was less in the late Precambrian, and that shells were developed to shield the soft-bodied creatures from radiation. Others think that shells afforded protection against predators, the latter appearing for the first time in late Precambrian times. Climatic and chemical changes in sea water have also been suggested to account for what palaeontologists accept as one of the major mysteries of the fossil record.

Whatever the reason, the advent of skeletons is reflected in the increased record of past life present in the Cambrian period.

Charnia masoni, a problematic fossil from the Precambrian of the Charnwood Forest, Leicestershire, England. 21 cm (8·2 ins) long.

Medusina dawsoni, the doubtful cast of a Precambrian jellyfish from the Ediacara Hills, South Australia. 1·5 cm (0·6 ins) in diameter.

The Palaeozoic era

The glass sponge *Hydnoceras* was very abundant during the Devonian, particularly in western New York State, USA. Average length 13 cm (5 ins).

Of the various phyla set out on page 16, six, the Cyanophyta, Echinodermata, Protozoa, Porifera, Annelida and Coelenterata are recorded from Precambrian sediments. In the Cambrian no less than twelve phyla occur, the six mentioned above plus the Brachiopoda, Mollusca, Arthropoda, Bryozoa, Chordata and Hemichordata.

Of the eleven, several possess skeletal hard parts of diverse character and composition. The skeletons common to both plants and animals support and protect the soft tissues which perform all the functions essential to life. Chitinous, calcareous, siliceous and chitinophosphatic hard parts are all known from the Lower Cambrian.

In plants, calcareous traces remain common to the blue-green algae in the form of stromatolites. Excellent examples of stromatolitic reefs are found in New York State and Canada. The laminated structure of stromatolites, like that of modern blue-green algae, is formed by the trapping of sediments by dense coverings of algae in the intertidal zone. Algae similar to modern seaweeds have been reported from Cambrian sediments. If correct, it is probable that these plants lived in slightly deeper waters than those involved in the development of stromatolites. New forms of animal life were also abundant in Cambrian seas.

Simple and complex animals existed together in these early Cambrian times, different organisms performing various roles within a given community. Doubtful records of single-celled protozoans, in the form of agglutinated tests of early foraminiferans and the siliceous tests of radiolarians, have been made from Early Cambrian sediments. Agglutinated tests are formed by the cementing of sedimentary particles to the outer membrane of the organism. Such tests are usually associated with foraminiferids that live in shallow waters. Some genera, like *Textularia*, cement quartz grains to their tests in cold water areas and calcareous grains in warm tropical seas. The complex tests of the Radiolaria are mainly external structures, the siliceous skeleton having the form of a complex lattice work. The radiolarians are predominantly planktonic floaters, their skeletons sinking to the sea-floor to form thick oozes at depths up to 4 500 metres and beyond. Radiolarian oozes are unknown in the Cambrian but the records of both Radiolaria and agglutinated Foraminiferida provide information about ancient ways of life.

Early foraminiferids are also simple in structure, single chambers or small clusters of chambers being the height of complexity. Throughout the Cambrian and Ordovician the agglutinated varieties dominate. Calcareous forms which appear during the Ordovician became abundant during the Silurian period, and undergo a significant expansion during Devonian times. In the Carboniferous, foraminiferids occur in such abundance as to be of importance in local sedimentary environments as rock formers. Rather complex structural types appear during the Late Palaeozoic with two superfamilies, the Endothyracea and Fusulinacea, being of particular importance in the Carboniferous and Permian periods. Of the two the fusulines evolve very rapidly, the size, shape and internal structure of the tests changing so quickly that the vast number of short-lived species that result are ideal zone or index fossils for the subdivision of the Upper Carboniferous and Permian periods.

Sponges (Porifera) are amongst the simplest multicellular organisms discovered in the Cambrian and like the protozoans can be traced back to Precambrian times. They are exclusively marine animals, most of which remain attached to the substrate throughout their life. Simple sponges are characterized by the presence of pores through which water carrying food and oxygen passes into a large internal chamber. The outer wall, pores and chamber lining are composed of living cells, some of which resemble individual protozoans. These cells are organized into distinct regions and the whole is supported by a skeleton made up of spicules. As fossils these are found often as disarticulated cumulations and only when they become fused to give a rigid structure are they of any real value.

The earliest sponges were probably simple, sac-like structures but later forms had increased surface areas and improved systems of water circulation. Different types of sponge structures evolved during the Precambrian, and Cambrian sponge communities exhibited a wide variety of different genera and species. The majority lived a static existence fixed to the sea-floor, either as individuals or in large clusters. Some, particularly in later Cambrian times, may have attached themselves to floating algal masses and drifted with the currents of the ancient oceans.

Rigid or interlocked frameworks of spicules are known in number from the Burgess Shales of British Columbia, Canada. These shales, laid down in calm conditions, offered an ideal situation for the preservation of a Middle Cambrian community. Many other animals are preserved alongside the early representatives of the siliceous sponges. Siliceous forms dominate Lower Palaeozoic sponge faunas and in general suggest rather deep, cold-water environments.

In the Devonian period, calcareous sponges appear for the first time. Initially they are overshadowed by the Hyalospongea, or 'glass sponges', which are found in large colonies in areas such as the eastern region of New York State, where forms like *Hydnoceras* were particularly abundant. The Calcispongea increase in number and variety throughout the Devonian and become important reef builders throughout New Mexico and Texas in Carboniferous times. Unlike their siliceous relatives, calcareous sponges thrive in warm, shallow-water environments.

Other organisms, more complex than the sponges, evolved during the Precambrian. Some, like the Archaeocyatha and Coelenterata, represent the next rungs on the evolutionary ladder, the similarity of the archaeocyathids to the sponges earning them the name of pleosponges in certain texts.

Restricted to rocks of Cambrian age, the mainly conical-shaped archaeocyathids have porous walls and a large central cavity. Unlike the earliest sponges, however, they have a calcareous skeleton. Single- or double-walled specimens have been discovered. No record of soft parts has survived but it is probable that the animal shared features of both the sponges and the coelenterates. The base of the skeleton is rather dense and covered with thin 'root-like' structures, that suggest a fixed mode of life.

Archaeocyathids are found throughout the world during the Lower and Middle Cambrian. Fossils from America, Australia, China, North Africa, northern France and Siberia illustrate the wide geographical distribution of the group, whilst outcrops extending for over 640 kilometres (400 miles) in Australia, reveal the importance of archaeocyathids as 'reef garden' dwellers. It is thought that they disliked muddy conditions, and from frequent associated discoveries, that they lived in symbiotic relationships with trilobites such as olenellids. Similar relationships exist today where small reef-dwelling fish live inside deadly stinging corals. In return for the womb-like protection of the coral's tentacles and central cavity, the fish cleans away unwanted materials.

The presence of the archaeocyathids in the Cambrian is problematic in that they have no apparent ancestors or, in younger rocks, any descendants.

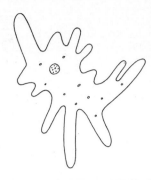

Amoeba, a foraminiferid which lacks a shell.

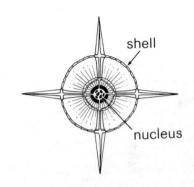

Radiolarian, a foraminiferid with a shell in the form of a siliceous lattice.

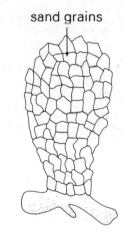

Agglutinated foraminiferid.

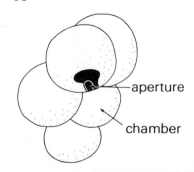

Globigerina, a calcareous foraminiferid.

Protozoa

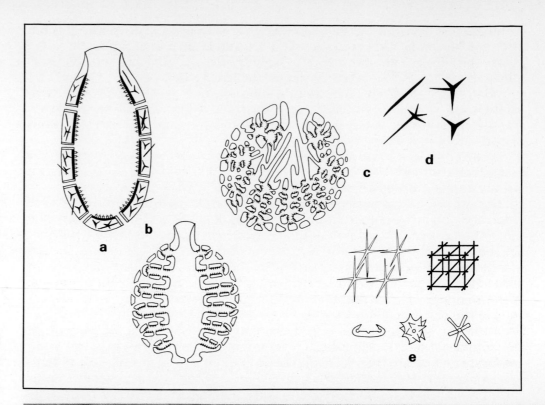

In simple sponges (a) the body is vase-like whilst in more complex types (b and c) the body wall is folded to form many subsidiary chambers. The majority of sponges have a skeleton of spicules (d and e). a Simple asconoid sponge; b syconoid sponge; c complex leuconoid sponge; d megascleres (large spicules); e fused and unfused six-rayed spicules, with three microscleres (small spicules).

Sponge-like organisms, the archaeocyathids flourished in Lower and Middle Cambrian times. They are often found in association with trilobites. 9 mm (0·4 ins) in diameter.

Far right
The tabulate coral *Halysites catenularius* (Ordovician-Silurian) is commonly known as the chain coral. Wenlock limestone, Dudley, England. Single corallite 1·2 mm (0·5 ins) wide.

At times they have been associated with both sponges and coelenterates but at the present they are considered as a phylum in their own right.

Like the sponges and protozoans, the history of the coelenterates began in Precambrian times. The word coelenterate means 'hollow gut'. The animals included in this phylum have well developed body tissues and radial symmetry is also an important characteristic of the group. As with the sponges, the coelenterates lack body organs and have no anus and no central nervous or circulatory systems.

All coelenterates, living and fossil, are aquatic invertebrates, the adult form being highly varied. The vast majority of recorded species are colonial, but solitary forms are known throughout the fossil record and free-swimming, floating and bottom-dwelling forms are common to both individual and colonial types. Two basic structural types of adult coelenterate are known, one the umbrella-shaped medusa found in jellyfish, the other the sac-like polyp of the sea-anemones and corals. The latter has the mouth on the dorsal surface surrounded by a crown of tentacles. In the medusa, the tentacles occur around the margin of the body and the mouth is central on the ventral surface. In the majority of coelenterates the tentacles possess sting cells which are used in the capture of food.

Of the two structural types mentioned above the medusae of the jellyfish are amongst the first recorded coelenterates, their presence being listed amongst the fauna of the Ediacara community of Australia. The lack of hard parts makes the preservation of the jellyfish or scyphozoans a rather fortuitous affair, but, in the exceptional conditions that prevailed during the deposition of the Burgess shales, representatives of four of the five known orders of jellyfish are preserved as impressions.

Further, rather isolated, recordings of jellyfish have been made throughout the Palaeozoic but the geological record of these coelenterates is overshadowed by the appearance of polypoid forms with hard parts during the Ordovician period. These are the Palaeozoic corals, the soft parts of which resemble a sea-anemone, evidence for this form of the polyp being preserved in the form of the calcareous skeletons of the Palaeozoic tabulate (Tabulata) and wrinkled corals (Rugosa).

The tabulate corals are all colonial, the skeleton of the colony being called the corallum. Individual tubes or corallites make up the colony, and according to species may be polygonal, circular or elliptical in shape. Within the walls of each tube an anemone-like polyp rests on a horizontal calcareous plate, the tentacles trapping microscopic organisms for food. The individual corallites are always in contact with others, the packing ranging from loose to compressed or massive. Tabulate corals are placed with wrinkled corals and modern hexacorals in the class Anthozoa.

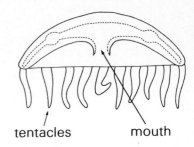

Medusoid type.

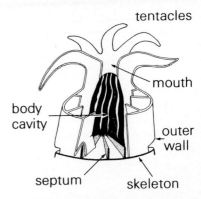

Polypoid type. Section to show structure of animal (polyp) and relationship of soft parts to skeleton.

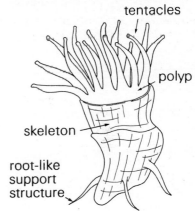

Suggested relationship of animal (polyp) to its skeleton in the wrinkled corals (Rugosa).

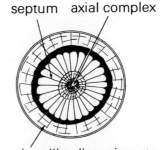

The cross-section of a coral skeleton.

Medusoid and polypoid types of coelenterate animals.

27

The rugose coral *Calceola sandalina*, Devonian, Belgium. Popularly known as the slipper coral, *Calceola* probably lived in quite strong currents. Width of corallite 3 cm (1·2 ins).

Tabulate corals such as *Syringopora reticulata*, (Lower Carboniferous, Kendal, England) were important reef builders during the Palaeozoic era. Height of colony 6 cm (2·3 ins).

Tabulate corals first appear in limited numbers in the early Middle Ordovician, small colonies being discovered in sediments of the Appalachian Mountains of the eastern states of North America. Individual variation increases throughout the Ordovician, and within a comparatively short period of time the tabulate corals had spread throughout the world. Reefs are common throughout the Palaeozoic history of the tabulates, and these corals play an important role in the building of these complex communities. Tabulate corals reached their acme in Siluro-Devonian times, when reef growth, associated with the calcareous sediments deposited in warm shelf-sea environments, was of particular importance. The tabulates, although limited in the variety of species recorded, were still common in the Carboniferous and Lower Permian, but, like their distant cousins the Rugosa, became extinct by the end of the Permian period and the Palaeozoic era.

The early history of the rugose corals was overshadowed by the development of the tabulates. At first the wrinkled corals were represented by small, solitary, horn-shaped structures such as *Lambeophyllum* and *Streptelasma*, from the Middle Ordovician of North America. Unlike the tabulate corals, these solitary corallites were characterized by strong septa, a feature shared by the majority of the rugose corals.

By the Silurian, colonial rugose corals had appeared in numbers and, like the tabulates, were important reef builders. It is more than likely that colonial rugose corals developed from solitary types. Asexual budding probably accounts for the development of the colonies, and the appearance of reefs during the Silurian may be associated with a sudden increase in this form of reproduction. Ideal environmental conditions such as warm, clear waters and abundant food stocks may have assisted in the expansion of the group at this time.

Solitary forms change considerably throughout the Ordovician and Silurian periods. Evolutionary changes take place in several stocks; the simple disc-like or horn corals of early times giving rise to elongate and cylindrical forms. Colonial corals also follow a number of trends, one being the change from those with a loose tubular form to colonies where the walls of individual units disappear and the septa become confluent over the surface of the colony.

Throughout the Upper Palaeozoic, the history of the rugose corals was marked by a series of evolutionary troughs and peaks. One peak took place during the Lower and Middle Devonian some 395–370 million years ago, but, unlike a similar event in the Lower Silurian noted for the occurrence of new morphological features, was characterized by a wealth of new families, genera and species. It is possible that ideal conditions were again responsible for this evolutionary burst. Rugose corals such as *Hexagonaria* and *Heliophyllum* are abundant in the Devonian sediments of Europe, Morocco and North America. In the Carboniferous, the number of rugose genera declines throughout the world, but in Europe a large number of genera exist which are as yet unknown in the United States. These are of sufficient importance in the Lower Carboniferous to be used as zone fossils.

Lithostrotion portlocki (Lower Carboniferous, Bristol, England), a colonial, rugose coral. Single corallite 7·5 mm (0·29 in) wide.

Naos pagoda from the Silurian of Arctic America is a representative of the rugose (wrinkled) corals. 8 cm (3·1 ins) in diameter.

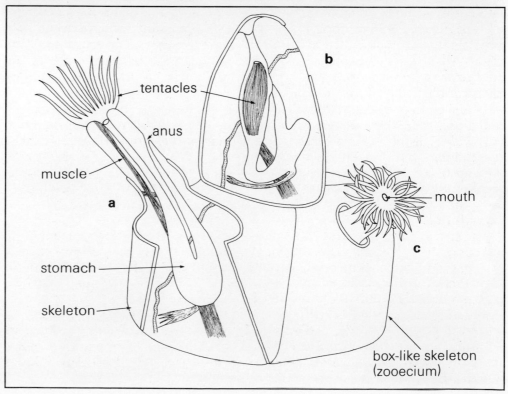

tentacles

anus

muscle

mouth

stomach

skeleton

box-like skeleton (zooecium)

b

a

c

A bryozoan colony sectioned to show form of skeleton and the nature of individual animals. **a** Section of animal (zooid) and skeleton in feeding position; **b** section of animal and skeleton in retracted position; **c** animal in extended position.

Heliophyllum agassizi, a solitary wrinkled coral from the Middle Devonian of Falls, Ohio, USA. 7·5 cm (2·9 ins) in diameter.

Morphologically, a great transformation took place between the Devonian and Carboniferous rugose corals. Many genera of the latter period developed solid or complex, web-like structures in the central area of the coral skeleton. The development of structures such as these, throughout their history, indicates that the rugose corals were continuously modifying their skeleton in response to changes in environment or mode of life. In the Permian the Rugosa, like their tabulate cousins, disappeared from the seas of the world.

Other organisms, often of greater importance, aided the rugose and tabulate corals in the building of reefs. Algae, stromatoporoids and bryozoans are all recognized as major contributors to reef binding and construction during the Palaeozoic era. The Stromatoporoidea are thought to be relatives of the corals, their massive, lamellate or branching colonies being referred to the Hydrozoa, a group of polypoid coelenterates, recent examples of which are *Hydra* and the millepores of coral reefs.

The skeletons of the stromatoporoids are calcareous, the first specimens being recorded from rocks of Cambrian age. By Ordovician times the group was widely distributed and during the Silurian and Devonian became major reef builders. It is possible to infer from the type of stromatoporoid present in a given bed the conditions that prevailed during its lifetime. Massive forms would indicate shallow, turbulent zones and delicate branched species, calm off-shore conditions. In the Upper Palaeozoic the importance of the stromatoporoids decreases.

The stromatoporoids and Palaeozoic corals are placed in the Coelenterata, indicating that they are lower on the evolutionary ladder than the Bryozoa or sea-mats. Individually the bryozoans are tiny, most being less than 1 millimetre (0·04 inches) in length, but in colonies many thousands of individuals may give rise to a skeletal frame over 60 centimetres (23 inches) across.

The individual animal is rather polypoid in character, having tentacles and a sac-like body. Internally, however, each animal has a well-defined digestive tract and a number of strong muscles. Many bryozoans secrete a mineralized skeleton, each minute animal being surrounded, in part or fully, by a series of calcareous walls.

Permian wrinkled corals from the Glass Mountains, Texas, USA. 3·5 cm (1·4 ins) long.

The reef-building stromatoporoids of the Palaeozoic era were characterized by the presence of dome-like structures and radiating canal systems. 4 cm (1·6 ins) long.

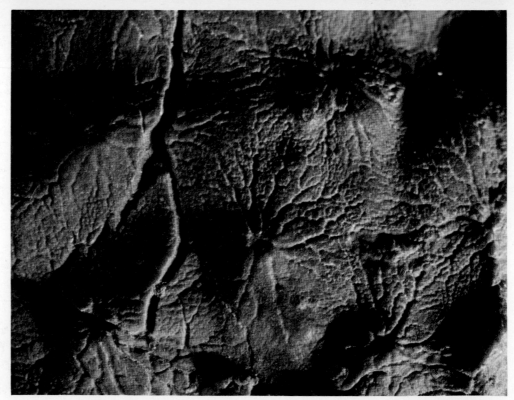

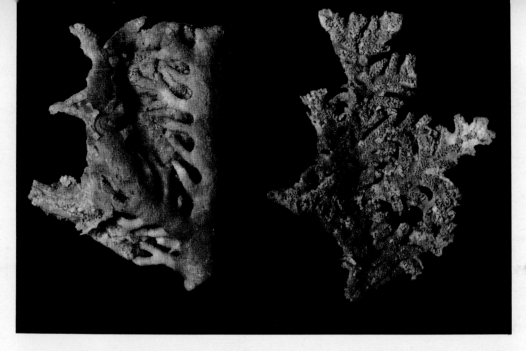

Permian bryozoans from the Glass Mountains, Texas, USA. Specimen on left 4 cm (1·6 ins) long.

Doubtful fossil bryozoans are recorded from the Cambrian, but by the late Middle Ordovician the picture has changed dramatically, over 100 genera having been described from this period of time. Mostly the Ordovician genera are members of the stony (Trepostomata) or hidden-mouthed bryozoans (Cryptostomata), two of the five orders belonging to the class Gymnolaemata. Four orders, the Ctenostomata, Cyclostomata, Trepostomata and Crypto- stomata, arise in late Cambrian/early Ordovician times. Members of these orders flourish throughout the Palaeozoic.

Generally, the ctenostomes are rather delicate thread-like bryozoans, many growing over the surface of pebbles or shells on the sea-floor. Delicate, cylindrical, twig-like cyclostomes are known from Palaeozoic rocks, but other colonies are globular or irregularly massive. Many 'stony' trepostomes, as their common name suggests, form rather massive colonies, but again branching, globular or sheet-like forms are common. In life these colonies may have been brightly coloured, the form and beauty of Palaeozoic reefs resembling that of reefs found today in the Pacific and Indian Oceans.

Some of the more massive bryozoans resemble the tabulate corals, but others like *Fenestella* and *Archimedes*, which grows out from a screw-like axis, are quite distinctive.

Of the organisms so far mentioned the majority can be regarded as rather simple. Many are involved in reef building and, apart from the radiolarians and medusoid jellyfish, live or lived on the sea-floor. Throughout the Palaeozoic more complex organisms evolve to fill different ecological niches. Included in these are the various invertebrate phyla noted earlier (Brachiopoda, Arthro- poda, Annelida, Mollusca and Echinodermata), together with the higher plants and chordates.

Of these groups, the brachiopods show certain anatomical links with the bryozoans. The external skeleton, however, is very different, the soft parts of a brachiopod being enclosed in a bivalved chitinophosphatic or calcareous shell. Each valve is equilaterally symmetrical, but usually the valves are different in size and shape. The ventral valve is the larger of the two, the differ- ence between the valves helping to separate the brachiopods from the bivalve molluscs.

The composition of the brachiopod shell is important in its classification. The brachiopods are divided into two major groups, the Inarticulata, which encompasses brachiopods with mainly chitinophosphatic shells, and the Articulata, whose shells are exclusively calcareous. Anatomically, the inarticulates have an anus, whilst the articulates have not. Within the valves the animals are similar, a body covering (mantle) enclosing the various organs. Structural differences between the shells of the two groups result in a greater number of muscles present in the inarticulates. The muscles, for opening and closing the valves, compensate for the lack of a well-developed hinge line, which is present in the vast majority of articulates.

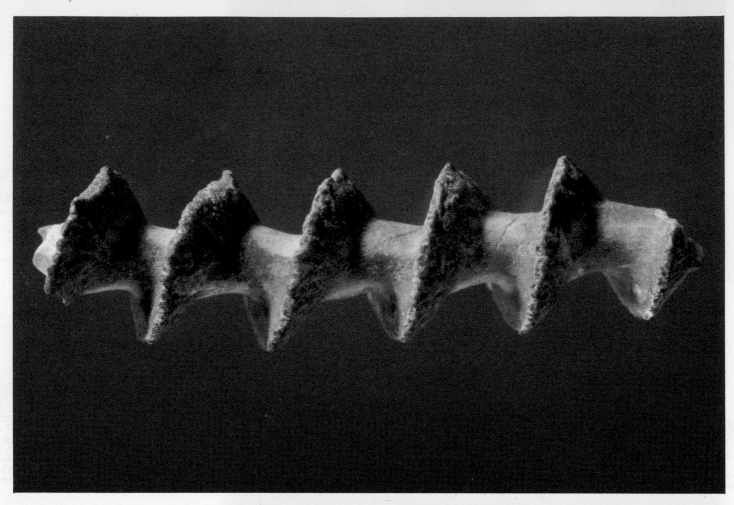

Archimedes (Lower Carboniferous, Alabama, USA) is a screw-shaped bryozoan. Length of axis 4·8 cm (1·9 ins).

Fenestella, the 'little windows' bryozoan, from the Permian, Glass Mountains, Texas, USA. 3 cm (1·2 ins) long.

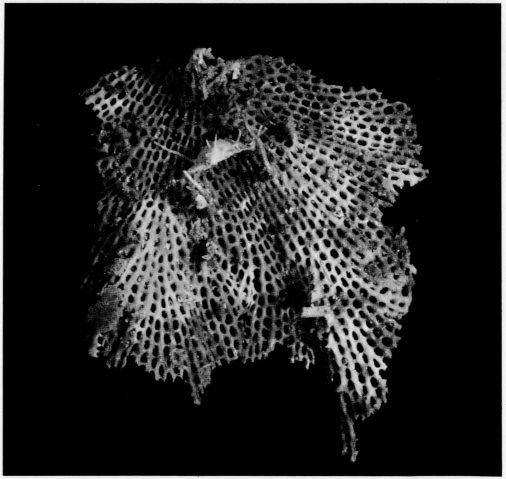

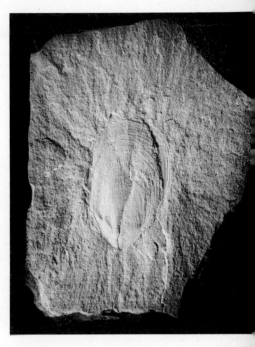

The inarticulate brachiopods are the simplest forms and one would expect them to pre-date their more complex relatives in the geological record, which they do. The first inarticulates are found in Lower Cambrian sediments. In certain cases the genera found in these rocks illustrate the very conservative nature of the group, for a few remain almost unchanged to the present day. Some Palaeozoic inarticulates adopt a burrowing mode of life whilst others attach themselves to the sea-floor or to other organisms. By direct comparison it is possible to establish that long range forms such as craniids (Ordovician – Recent) fed and obtained oxygen from two lateral water currents, and expelled waste by means of a stream of water directed along the mid-line of the shell. Calcareous representatives of the inarticulates arose early in the history of the group, but too late to be considered ancestral to the articulates, the earliest representatives of which are found in Lower Cambrian rocks.

Articulates, characterized by the presence of a hinge line with teeth and sockets, flourished during the Palaeozoic era, five out of six orders appearing by the end of the Ordovician. Cambrian and early Ordovician brachiopod communities are dominated by the inarticulates, orthids and pentamerids. Each species referred to these groups is to some degree noted for various physical features that reflect its mode of life, and palaeontologists use these features together with important field evidence to reconstruct past communities. The orthids and pentamerids were essentially sea-floor dwellers, the majority either resting on or attached to the substrate. Normally the orthids are characterized by a long hinge line and the biconvex shells are usually finely ornamented, characters shared in part by the early pentamerids. Both groups probably inhabited rather deep water during the earliest part of the Palaeozoic.

During the Ordovician the rather flat, broad, articulate brachiopods, strophomenids, increased in importance and millions of individuals have been found in Middle and Upper Ordovician rocks. Like the orthids many were sea-floor dwellers but others, again like the orthids, floated around attached to seaweed. The importance of the strophomenids continued in Silurian times, various genera being used in the evaluation of communities. Palaeontologists at work in various parts of the world have established the relationship of the communities to environment and correlated communities with depth.

Inarticulate brachiopods can be traced back over 500 million years, the similarity between *Lingulella* (**above**, 2·2 cm (0·9 in) long), from the Ordovician, and its Recent relative *Lingula* (**above left**, 2·7 cm (1 in) long) being rather obvious.

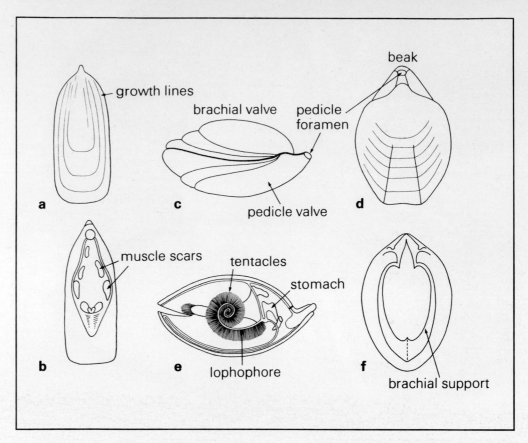

Lingula: **a** external view;
b internal view to show areas
of muscle attachment.
Terebratulid: **c** lateral view to
show arrangement of valves;
d view of brachial valve and
pedicle foramen; **e** longitudinal
section to show feeding organ
(lophophore); **f** internal view
of brachial valve showing
support structure.

During the Silurian and Devonian many pentamerids and a branch of the strophomenids, the productids, exhibit an overall increase in size. Their thick, often coarsely ribbed or spinose shell indicates adaptations to shallow, turbulent environments. In the Carboniferous, the productids in particular are very abundant and some develop long, rather hair-like spines which help to stabilize them in the face of quite strong current action. Others, like *Gigantoproductus*, although short-lived grew to enormous size, resembling the great clams of today. During the Permian, some productids, like *Prorichtofenia*, were rather like corals, individuals growing in tightly packed groups within reef masses.

From bite marks present on the valves of various Palaeozoic productids, it would appear that they acted as a food source for large, predatory nautiloids.

The order Rhynchonellida is characterized by small, biconvex shells, many of which are plicated (corrugated). The hinge line is short and curved and the shell has a pointed beak. The order, although well represented throughout the Palaeozoic, is found in greater numbers in the Mesozoic. The shape of a rhynchonellid shell often reveals evidence of its mode of life. The trilobed shell suggests well-developed feeding currents, and strong ribbing, a life in shallow waters. A number of distinctive species appear in the Devonian and Carboniferous which have limited ranges in geological time, but wide geographical distributions, making them ideal index fossils.

The orders noted above have a variety of internal supports for the lophophore, the internal appendage which creates the water currents within the shell. It would appear that the evolution of these structures is initially one of an increase in length, followed by an increase in complexity. The specialized support structures of the Terebratulida and Spiriferida give strength to this hypothesis.

In the majority of spiriferids the brachial support is well developed, showing considerable change during the Palaeozoic era. As the name suggests, most spiriferids have a spiral support structure. The shell is usually biconvex and the hinge line often the widest measurement. Normally, the shells are marked by radial ribs and many are plicate. The form of the shell and the nature of the spiral support structure suggests that the spiriferids had also developed very strong feeding currents.

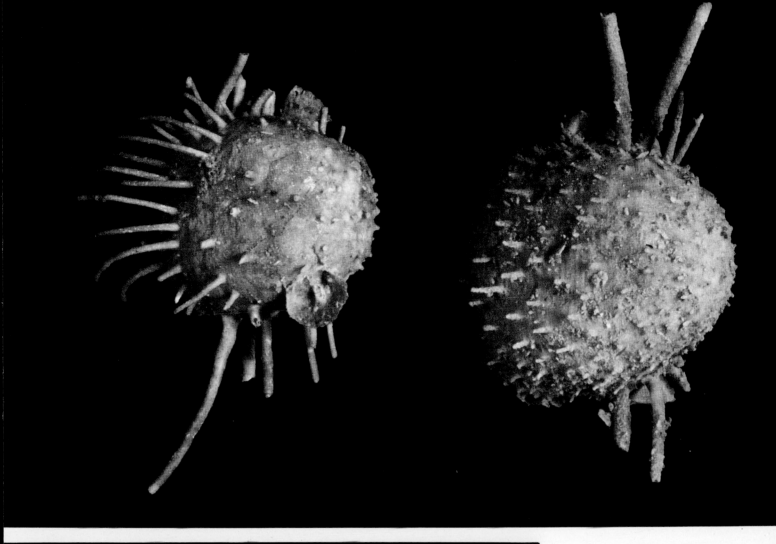

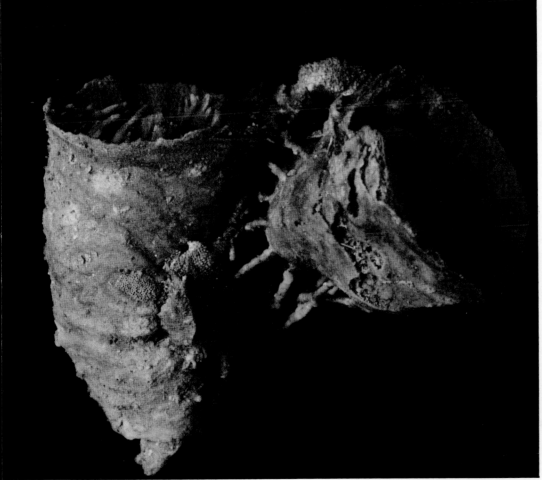

Productid brachiopods are
amongst the most spectacular
organisms found in Upper
Palaeozoic rocks. Specimens
from Permian, Glass
Mountains, Texas, USA.
Specimen on left 4·5 cm
(1·8 ins) long.

Coral-like brachiopods from
the Permian, Glass Mountains,
Texas, USA. Specimen on left
4·5 cm (1·8 ins) long.

The oldest spiriferids, the Atrypacea, occurring in the Ordovician, are rather rounded in form with a curved hinge line. *Atrypa* is a prominent representative of Silurian and Devonian communities and in some cases, individuals are seen to change their overall shape as a response to the rather soft nature of the sediment, a larger, broader shell preventing the organism from sinking into the fine mud on which it lived.

The more typical, often more spectacular spiriferids belong to the Spiriferacea, various species serving as ideal zone fossils in the stratigraphic analysis of Devonian rocks, of which *Cyrtospirifer* and *Mucrospirifer* are important representatives. Throughout the Carboniferous and into the Permian, a variety of spiriferids maintain the dominance of the group within Upper Palaeozoic faunas. The morphology of the group suggests that the majority of forms inhabited shelf-sea and deltaic environments. In some communities a mutually beneficial relationship is recognized between *Spirifer* and the tabulate coral *Aulopora*.

Unlike the more specialized spiriferids, the Terebratulida are somewhat pear-shaped. Their shells are usually smooth and a large pedicle opening (foramen) is present in many species. A large pedicle, for attachment, would suggest that fossil species like their modern counterparts occupied shallow water. Terebratulids are not very common in Palaeozoic sediments, but unlike the orthids, spiriferids, pentamerids and strophomenids, they survive into and flourish during the Mesozoic era. The others either disappear or survive in a limited capacity, Mesozoic brachiopods consisting mainly of terebratulids and rhynchonellids.

Of the remaining invertebrate phyla, the Arthropoda occupy a special position in the hearts of palaeontologists. The thrill of finding a trilobite or crab ranks highly amongst the rewards for devoted fieldwork. The arthropods are literally the 'joint-foot' animals, the body being segmented and covered by an external skeleton composed of chitin. In some groups the main layer of the skeleton is impregnated with calcium carbonate and calcium phosphate, which strengthen the hard parts and make the arthropods functionally superior to many other groups. Many are aquatic, particularly those of the early Palaeozoic, while others invade the land and many become active flyers. Over 700,000 species of arthropod have been recorded, accounting for three-quarters of all known animal life. Five subphyla exist which are divided into eighteen classes and of these, sixteen originated during the Palaeozoic.

Spiriferid brachiopods, like *Brachyspirifer* are important in the geological subdivision of the Devonian period. 4 cm (1·6 ins) wide.

Amongst the articulate brachiopods of the Silurian, *Leptàena* was abundant and worldwide in its distribution. 3·9 cm (1·5 ins) wide.

During geological times, the fossils of plants and animals can be deformed by the folding or movement of strata. These are deformed brachiopods, *Dinorthis multiplicata*, Ordovician, North Wales. Width of rock sample 44 cm (17 ins).

The appearance of highly organized animals, such as the trilobites in the Lower Cambrian, suggests a long period of evolution in earlier times. This theory is supported by the fact that a variety of fossil arthropods other than the trilobites are abundant in the Burgess shales. The character of the related phylum Parathropoda would suggest a link between them, the arthropods and the soft-bodied annelid worms, traces of which have been found in the Proterozoic rocks.

The worms, as stated, are soft-bodied, the parathropods encased in a flexible chitinous cuticle and the Trilobitomorpha in a mineralized external skeleton.

The subphylum Trilobitomorpha is divided into the Trilobita and several other groups, of which the Merostomoidea is possibly the most important. They are well represented in the Burgess fauna and are possibly ancestral to the king crabs and eurypterids. Like the trilobites, the merostomoid body segments are divided into well-defined regions, a trilobate dorsal shield covering the front part of the body. In the trilobites the body is divided both longitudinally and transversely, the body being divided into head (cephalon), thorax and pygidium. Ventrally, five pairs of jointed limbs are attached to the cephalon, one pair to each of the thoracic segments and a pair to each of the larger fused 'tail' segments. The vast majority of these are identical but the first pair function as antennae.

Little is known of the internal structures of trilobites, although markings suggest the form of various soft organs and indicate areas of muscle attachment. The general organization of the trilobites places them amongst the

The discovery of trilobite limbs is a rather rare occurence. The specimen of *Olenoides* pictured here exhibits both limbs and antennae. 6·4 cm (2·5 ins) long.

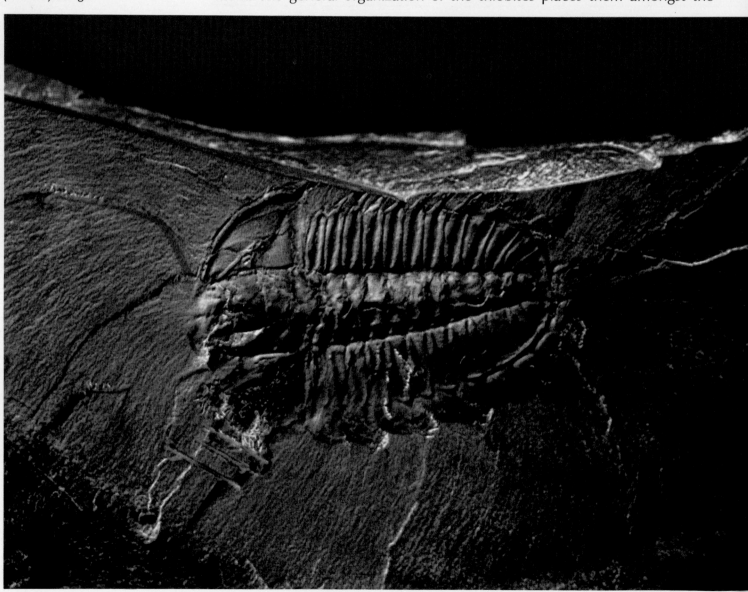

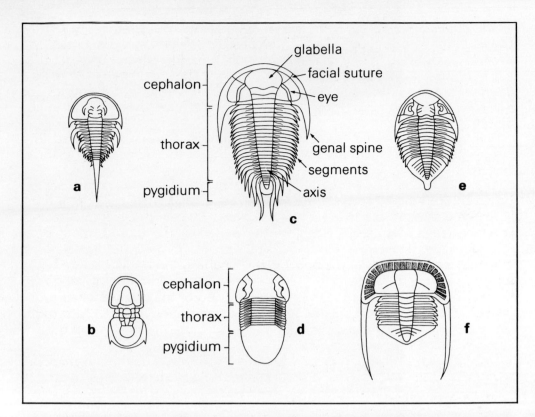

glabella
facial suture
eye

cephalon

thorax

pygidium

genal spine
segments
axis

a

c

e

cephalon

thorax

pygidium

b

d

f

more advanced invertebrates. The presence of compound eyes in many species illustrates the evolutionary progress that has taken place since the appearance of the humble single-celled organism.

The first trilobites appear in the Lower Cambrian, with such diverse forms as *Agnostus, Eodiscus* and *Paedeumias* indicating Precambrian roots. Facial sutures which facilitate the shedding of the skeleton during growth are useful in the classification of trilobites. In the case of the above-mentioned genera, a suture is lacking in *Agnostus* and *Eodiscus* but present in *Paedeumias*. The marginal structure of *Paedeumias,* sometimes termed protoparian, would appear to be ancestral to later structural types termed proparian and opisthoparian. In these the sutures occur on the dorsal surface, whilst in the hypoparians the suture line is again marginal.

The protoparian trilobites are relatively short-lived, lacking representation in sediments of middle Cambrian age. Of the other groups, the eodiscids die out by the start of the Upper Cambrian and the agnostids at the end of the Lower Ordovician. Trilobites with opisthoparian sutures dominate Palaeozoic arthropod communities, with proparian and hypoparian being abundant constituents until late Devonian times.

Evolutionary trends among the trilobites are difficult to trace, but in general terms various lineages follow trends related to the form of the head shield, the number of thoracic segments, the fusion of pygidial segments and the development of spines.

These trends, coupled with the evolutionary development of the facial suture and the various changes that occur in the size and shape of the eye, are probably associated with adaptation to environment and mode of life. Most trilobites crawled over the sea-floor leaving behind traces of their movements and burrowing habits, but a few large 'tailed' or spinose forms were probably free-swimming or planktonic. Trilobite communities have proved useful in reconstructing the geography of early Palaeozoic times. In the absence of other fossils, trilobites are used for the subdivision of the Cambrian; *Olenellus* (protoparian), *Paradoxides* (opisthoparian) and *Olenus* (opisthoparian) being used as zone fossils for the Lower, Middle and Upper Cambrian, respectively. The importance of the trilobites both as index fossils and members of Palaeozoic communities diminishes after the Cambrian, although certain genera such as *Trinucleus* and *Phacops* are of considerable importance during the Ordovician and Devonian.

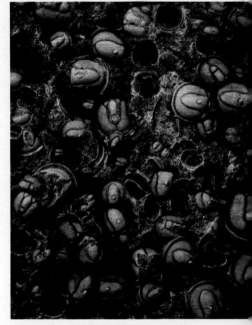

Moulds and casts of *Agnostus,* a tiny, blind trilobite of Cambrian age. Width of head shield 8 mm (0·3 ins).

Above
Trilobites such as *Tretaspis sortita* (Upper Ordovician, Girvan, Scotland) were blind, sea-floor dwellers. 3·3 cm (1·3 ins) long.

Above right
Ogygiocarella debuchi, an Ordovician trilobite. 4 cm (1·5 ins) long.

Right
The trilobite *Deiphon* (Wenlock limestone, Malvern, England) is thought to have been an open-sea surface dweller. 4·4 cm (1·75 ins) long.

Bumastus is an excellent example of an isopygous trilobite, in which the head and tail are of equal size. 4·9 cm (1·9 ins) long.

A trilobite *Calymene blumenbachi*, Middle Silurian, Wenlock limestone, Dudley, England. 5 cm (2 ins) long.

Eurypterus lacustris (Silurian, Buffalo, USA) and its euryp-terid brethren are often called the giant water scorpions. 14·5 cm (5·7 ins) long.

Like many other Palaeozoic orders, the trilobites became extinct by the end of the Permian period and the possible reasons for this are discussed at the end of this chapter. The extinction of the trilobites is by no means the end of the arthropods, for although the trilobites are the most common Palaeozoic forms others, less significant during this era, survive into the Mesozoic and Cainozoic.

The subphylum Chelicerata, for example, is represented in the Palaeozoic by the scorpions and spiders (class Arachnida); the king crabs and eurypterids (class Merostomata) and the extremely rare sea spiders (class Pycnogonida). Unlike the trilobites, the chelicerates have no antennae and the limbs situated in front of the mouth have pincers.

Of the three classes noted above, the merostomes are possibly the most important. The development of the king crabs begins in the Middle and Upper Cambrian with the appearance of the aglaspids. In the Silurian and Devonian other relatives, the synxiphosurans, occur, whilst the true limulids or king crabs arise in Devonian times.

The eurypterids are amongst the largest of known arthropods, with individuals reaching 3 metres (10 feet) in length. They range from the Ordovician to the Permian, reaching a significant peak during the Devonian period. Considerable argument over the mode of life of these spectacular organisms has failed to provide a satisfactory solution. In the Ordovician, eurypterid remains are found in association with marine sediments, but in the Silurian and Devonian they are linked with brackish water environments. One suggestion to account for this is that eurypterids spent their lives in fresh water and were, on death, carried downstream into the sea. However, the chances of the segmented external skeleton remaining articulated during transportation are small. No freshwater eurypterids have been discovered, nor have any been found in association with the normal shelf-sea invertebrates.

The streamlined form of the eurypterids, together with the oar-like nature of the last pair of limbs, suggests animals suited to an active lifestyle. Some may have been bottom crawlers and scavengers, others swimmers and predators.

Of the arachnids, both spiders and scorpions are represented in the Palaeozoic. The first scorpions appear in the Silurian and the form *Palaeophonus nuncius* has been described as the first land dweller. If this is true, then the event is of considerable significance and probably corresponds with the migration of vascular plants on to the land. Silurian scorpions have eight segments in front of the abdomen, whilst the true scorpions of the Carboniferous have seven. Palaeozoic scorpions have been found in Sweden, Scotland and North America.

Fossil spiders are rare, but can be traced back as far as the Devonian. For many millions of years they remained an insignificant part of the world's fauna, but in the Upper Carboniferous numerous new groups appeared, their occurrence being linked with the great swamps and forests of that time.

Of the other Palaeozoic arthropods the insects first appear in the Rhynie chert (Middle Devonian) of Scotland. Although specialized in many ways, these early forms are wingless. Relatives of the modern 'silverfish', the wingless fabric feeder, are found in Upper Carboniferous sediments.

Winged insects appear in the Upper Carboniferous of Europe, North America, Russia and Australia. Hundreds of different species found in various localities herald the appearance of a group which will expand dramatically and account for several hundred thousands of species within the animal kingdom. Cockroaches, beetles, grasshoppers, flies, lacewings and bugs all have their origin in late Palaeozoic times.

The Crustacea, known today mainly through the aquatic lobsters, crabs and crayfish, is represented in abundance in Palaeozoic rocks, by the subclass Ostracoda. These are tiny animals, living encased in a bivalve shell which opens for feeding and swimming. The animal is characterized by seven pairs of limbs. The abundance and variety of ostracods in past times make them useful guide fossils. Their size (0·5–20 millimetres, 0·02–0·08 inches), distribution and rapid evolution adds to their usefulness, and their occurrence in freshwater sediments, where foraminiferids are rare, makes them important stratigraphic tools. Like foraminiferids, they are important in the correlation of sedimentary formations drilled in the search for oil. Ostracods are found first in Lower Ordovician rocks.

Relatives of the ostracods such as barnacles, branchiopods, leaf shrimps and others, are fairly common in Palaeozoic sediments. The crabs, lobsters, shrimps and prawns appear after the Palaeozoic.

Of all Palaeozoic invertebrates, some of the largest and most spectacular are found in the phylum Mollusca, the classes of which embrace a vast variety of marine, freshwater and terrestrial organisms. Snails, clams, squids, tusk shells and chitons form a group characterized by bilateral symmetry, concentration of sensory organs in a head (not true of bivalves), and total or almost total lack of segmentation.

An ostracod, *Beyrichia*, Silurian, Welsh border. The left valves of the female (**left**) and male animals. Female 1·95 mm (0·07 ins) long; male 1·9 mm (0·07 ins) long.

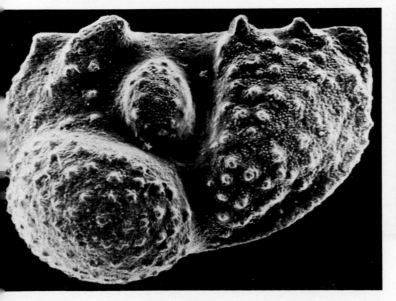

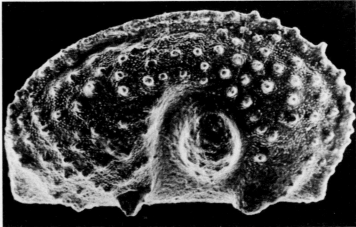

Molluscs are soft-bodied creatures, many of which protect themselves by means of an external shell. Some, probably more specialized, have internal support and stabilizing structures. The variety of shell form and structure make the molluscs very important tools in stratigraphy, and provide vital information about the mode of life and morphology of the mollusc. At the present time over 60,000 species of mollusc occupy a vast variety of ecological niches, and in many cases direct comparisons can be made between living and extinct forms.

The molluscs are divided into six classes, soft-part morphology, shell form and mode of life being used in the subdivision of the phylum.

The majority of species referred to the Mollusca are marine organisms, some being inhabitants of the sea bottom whilst others are free swimming. A number of bivalve and gastropod families live in fresh water and some gastropod genera dwell on land. The tusk shells are exclusively marine, and adopt an almost vertical position to burrow into sediments in tidal or deep-water zones. Geologically they are unimportant, although they may be abundant in certain strata. The chitons are also unimportant geologically and, like the tusk shells, are bottom dwellers, crawling slowly over the sea floor.

Gastropods, bivalves and cephalopods are the main molluscan stocks, each having prominent representation within Palaeozoic, Mesozoic and Cainozoic communities. In the 570 million years since the appearance of the first mollusc, the phylum has evolved successfully, with over 100,000 living species and many tens of thousands of fossil ones.

The gastropods first appear in the Lower Cambrian, their single-valved shells becoming progressively more abundant throughout Palaeozoic times. Individual families exhibit different geological distributions and many of the earliest and most primitive stocks which flourished in the Ordovician and Silurian die out by the end of the Palaeozoic, their niches being occupied by more specialized forms such as the pulmonates (with lungs).

The use of the electron-scan microscope has revealed the true beauty of minute animals such as the ostracod *Kellettina* (Carboniferous, Switzerland). 0·97 mm (0·04 ins) long.

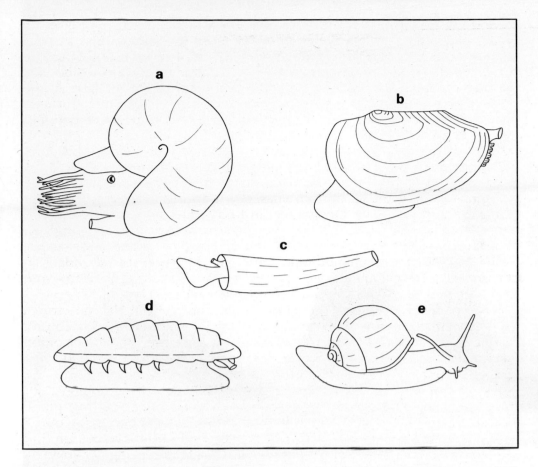

The Mollusca constitutes a large group of invertebrate animals. **a** Cephalopoda (ammonoids, belemnoids, octopuses, cuttlefish, squids); **b** Bivalvia (oysters, mussels, clams); **c** Scaphopoda (tusk shells); **d** Amphineura (chitons); **e** Gastropoda (slugs, whelks, snails).

Classification of the Mollusca

Amphineura (chitons), elongate, segmented shell consists of eight transverse units; large, broad muscular foot; two or more sets of gills. Geological range Ordovician – Recent.

Scaphopoda (tusk shells), shell unsegmented, elongate and tubular; foot rather pointed; no gills; head poorly developed. Geological range Silurian – Recent.

Cephalopoda (nautilus, squid, octopus), shell coiled planispirally, some forms naked, others with internal calcareous structures; head – foot tentacular with large eyes; either one or two pairs of gills – Dibranchiata (two gills), Tetrabranchiata (four gills). Geological range Cambrian – Recent.

Gastropoda (snails), shell seldom uncoiled or planispiral, mainly conical; foot large, broad and flat; gills or ctenidia paired, single or absent, lungs in terrestrial forms; well-defined head with eyes and tentacles. Geological range Cambrian – Recent.

Bivalvia (clams), shells bivalve with dorsal hinge; foot compressed laterally; paired gills; no head. Geological range Ordovician – Recent.

Monoplacophora, primitive with single, cap-shaped dorsal valve. Geological range Lower Cambrian – Recent.

Simple, non-coiled and cap-shaped shells are amongst the earliest and most basic gastropod remains. The gastropods are thought to have evolved from a bilaterally symmetrical mollusc, and the non-coiled shell meets with the simplest of protective requirements. Later forms are coiled, some symmetrically in a plane (planispiral), and others excentrically in a conical spire (conispiral). A loss of symmetry related to the twisting of the body occurs in the vast majority of gastropods. This torsion brings the gills and anus into an

Pleurotomaria anglica, a turbinate gastropod from the Silurian. 5 cm (2 ins) long.

anterior position and was important in the success of the group as the head could be withdrawn easily and when inside the shell was protected by the thick foot and operculum (horny or horny-calcareous plate). The majority of gastropods are coiled to the right (dextral), and only a few to the left (sinistral). To avoid contamination by their own waste material, many early gastropods developed a sinus or deep slit in the aperture of the shell. In later forms the sinus and slit are replaced by a posterior groove, the gutter. The development of the gutter is linked with the presence of inhalent siphons and with an increase in general activity. The siphons are tube-like and in drawing clean water into the shell protect the gills against sand or mud.

Palaeozoic gastropods belong mainly to the Amphigastropoda and Prosobranchia. The former reach their acme in the Ordovician and Silurian, whilst the latter are important throughout the fossil record. In Lower Palaeozoic rocks evidence exists to suggest an association between certain gastropods and nautiloids.

Anthracopupa, one of the oldest air-breathing terrestrial snails, is found in the Upper Carboniferous rocks of North America.

Within the class Cephalopoda, the living genus *Nautilus* is the sole surviving member of a very important group of shell-bearing molluscs, the Nautiloidea (similar to *Nautilus*). These are the first cephalopods recorded from the fossil record and from them the extremely successful ammonoid stocks arose in the late Silurian.

The modern *Nautilus* is regarded as a primitive cephalopod. Its eyes and circulatory system are comparatively simple and the animal lacks suckers on the tentacles, and an ink gland. Unlike many fossil nautiloids, the living form lacks internal deposits of calcium carbonate and the tightly coiled shell is different from the straight, curved or loosely coiled shells of earlier forms.

True nautiloids from the Upper Cambrian exhibit a variety of small, curved or straight shells, which lack the internal deposits of many later groups. The size of the shells and the absence of deposits suggests that these molluscs experienced few problems in maintaining their feeding stance. In the Ordovician, numerous genera appear in which the siphuncle is characterized by the presence of tube and cone-shaped deposits. The shells of these forms are often much larger than their Cambrian ancestors and the deposits represent weights to counter an increase in buoyancy. Many of the early Ordovician nautiloids are referred to three orders, the Ellesmeroceratida, Endoceratida and Nautilida, and a large number lived as sea-bottom scavengers and predators. An increase in shell size and in the number of chambers would increase the buoyancy of the shell and in the case of a straight or slightly curved structure would lift the shell from the horizontal to the vertical position, and a change of this type would seriously affect its feeding habits and stability.

Throughout the Ordovician, numerous structural adaptations are seen in the nautiloids and their success is witnessed in increased numbers and a greater distribution. The general trend of size increase is seen in endoceratid stocks, some of which reached 4 metres (13 feet) in length. Chamber deposits appear in the Ordovician Actinoceratida and like those of the siphuncle have a hydrostatic function. Other nautiloids solved their problems in different ways, the Ascoceratida having larger chambers on the dorsal surface. The juvenile and mature areas of the ascoceratid shells are different and in many forms the gently curved early portion is discarded. The bulbous form of the adult shell is probably an indication of a change of life style in later years, an adult ascoceratid being an active, free-swimming predator.

In the Middle Ordovician the important orders, the Michelinoceratida and Oncoceratida appear. Coiled and curved nautiloids become increasingly important during the Upper Ordovician and Silurian. The form of their shell often indicates an increase in mobility and many lived in and around reef masses occupying the same niches as certain present-day fish. Silurian cephalopod faunas are dominated by the stubby oncoceratids which continue to thrive during the Devonian. The michelinoceratids give rise to the first of the ammonoids and only the michelinoceratids and a few genera from both the actinoceratids and nautilids survive into the Carboniferous period. Tightly coiled nautilids, similar to the living representative, flourish during this period and it is they that represent the class in Mesozoic faunas. Unlike their ammonoid descendants, the majority of nautiloids have simple septa (partitions

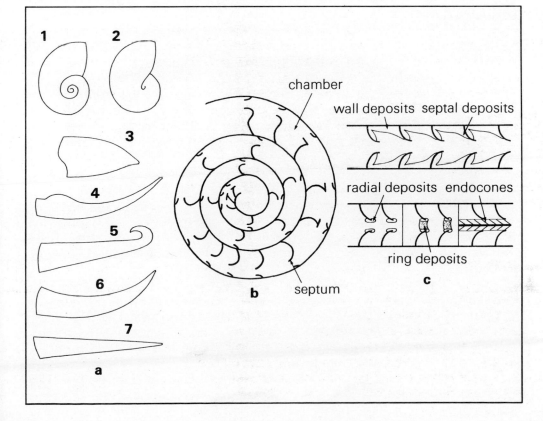

During the evolution of the nautiloids, numerous shell shapes appeared. Often, internal deposits were laid down to aid buoyancy problems. **a 1** evolute, **2** involute, **3** brevicone, **4** ascocone, blunt mature section, curved juvenile section, **5** slightly incurved, **6** slightly curved, **7** straight; **b** cross-section to show chambers and septa; **c** various internal deposits.

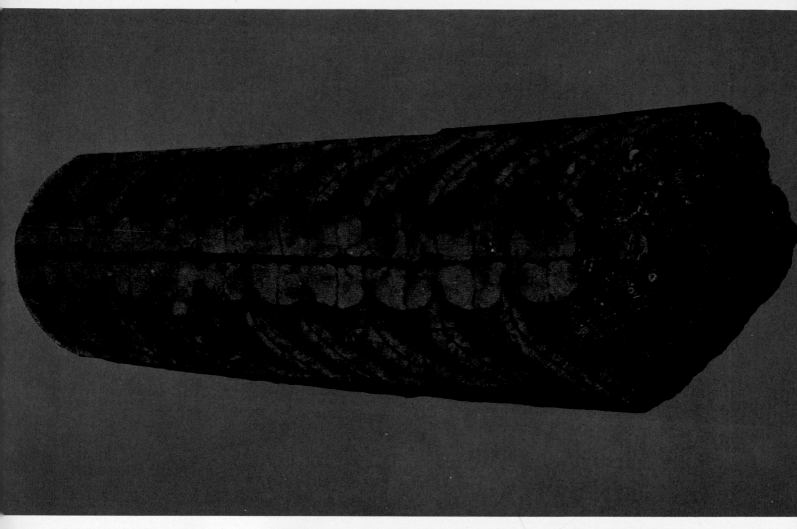

Many Palaeozoic nautiloids such as *Rayonnoceras* from the Lower Carboniferous are characterized by the presence of deposits in both chamber and siphonal areas. 17·7 cm (6·9 ins) long.

A silurian nautiloid, *Gomphoceras* from Bohemia. 4 cm (1·6 ins) long.

between chambers). A number, however, develop fluted septa and a marginal siphuncle; it is from these forms that the ammonoids arose. The genus *Bactrites* is thought to be the form directly linked with the evolution of the class Ammonoidea.

The first ammonoids appear in the Lower Devonian of West Germany. Only three genera represent the group at this time and all are characterized by the presence of a septal suture line which is folded into forward-directed saddles and receding troughs or lobes. This type of suture line is termed goniatitic and the goniatites undergo a dramatic evolutionary burst before the end of the Lower Devonian. From their Germanic cradle they spread out rapidly and have a world-wide distribution within a relatively short period of geological time. The vast majority of Devonian ammonoids appear to have been open-sea forms, being found in deep-water shales and off-reef lime-stones. The rapid evolution of species, linked with a wide geographical distribution, makes the goniatites ideal zone fossils throughout the Devonian. The major zones established in Germany are termed 'Stufen'. Early goniatites were loosely coiled or evolute (coiled, whorls touching), whilst later forms are mostly involute (coiled, outer whorls embracing inner ones). The majority are also typically goniatitic but one group, the clymenids, have the siphuncle located at the dorsal margin of the shell. At the close of the Devonian period, the clymenids die out, as do many of the goniatites. These extinctions are reflected in the fossil record by a restriction in distribution of the group as a whole. Those genera of goniatites that persist into the Carboniferous flourish in particular areas. Goniatites are important in the zonation of north-western European sediments. In the Upper Carboniferous a new group of ammonoids, the ceratites, originate from goniatitic stock. The numbers of ceratites increase throughout the remainder of the Palaeozoic, replacing the goniatites as the dominant group by the end of the Permian.

Ceratitic ammonoids have suture lines in which the lobes are subdivided into second-order folds. The fluting and folding of the septum probably enhanced the ability of these organisms to change their specific gravity, and consequently affected their mobility and habits. The ceratitic sutures become progressively more complex during the late Carboniferous and Permian. During the early part of the Permian the third and last suture type appears, the ammonitic type, in which both lobes and saddles are subdivided. Ammonites are of little importance during the Permian. Their contribution to various Mesozoic faunas is, however, unchallenged by any other invertebrate group.

Unlike their gastropod cousins, the bivalves have a skeleton that consists of two valves. These are hinged dorsally and enclose the soft parts of the animal. The bivalves are generally bilaterally symmetrical and, apart from lacking a well-defined head, have relatively simple circulatory, digestive and nervous systems. In many texts the group is called the Pelecypoda or 'hatchet feet', the fleshy, muscular foot projecting through the gape of the valves during movement or feeding. They are essentially 'primitive' molluscs, their development being restricted by the straightjacketing of the encompassing shell. The bivalves are bottom dwellers living mostly in shallow shelf-sea environments. A few genera are capable of swimming but most are sluggish crawlers, fixed bottom dwellers, burrowers or borers. The form of the shell varies considerably related to the mode of life of the animal. As with other groups, the classification of fossil species is difficult due to the lack of soft parts, and few pointers exist that help the palaeontologist in the reconstruction of the actual animal. In most forms a line, the pallial line, marks the line of attachment of the mantle musculature. This may be deflected posteriorly to form a sinus and indicate the area into which the siphons were retracted. The perfection of siphons in the early Mesozoic probably aided the expansion of the group. Elongate siphonal tubes are particularly characteristic of burrowing types. Prior to the Mesozoic expansion, the bivalves had increased in abundance and variety throughout the Palaeozoic. Of the eight recognized orders, six had appeared by the end of the era. Both marine and freshwater bivalves are included in these orders.

Bivalve faunas of the Ordovician include some of the most primitive forms known. These are characterized by rather thin shells and a poorly

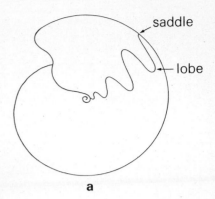

a

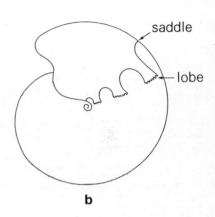

b

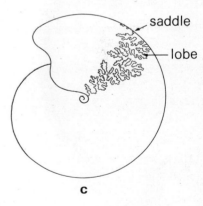

c

The suture lines of the ammonoids are important in their classification. **a** Goniatitic suture; **b** ceratitic suture; **c** ammonitic suture.

Right
A polished limestone section showing *Gastrioceras carbonarium* and other goniatites. Specimen from the Coal Measures, Upper Foot Mine, Littleborough, England. 2 cm (0·8 ins) in diameter.

Below
An early ammonoid *Manticoceras* from the Devonian of the Eifel region, Germany. 1·1 cm (0·4 ins) in diameter.

Below right
The ammonite *Gastrioceras* is used by palaeontologists in the geological subdivision of the Coal Measures, Upper Carboniferous. 3 cm (1·2 ins) in diameter.

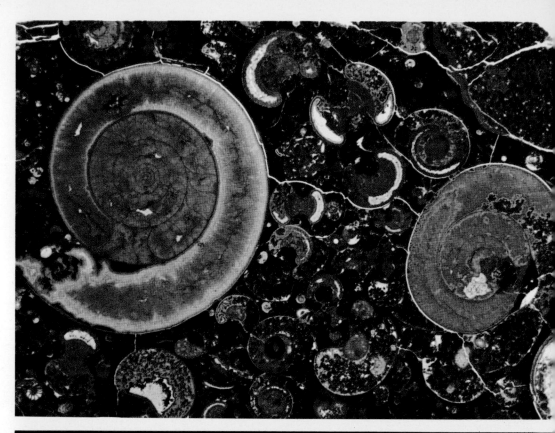

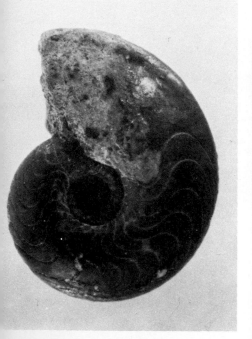

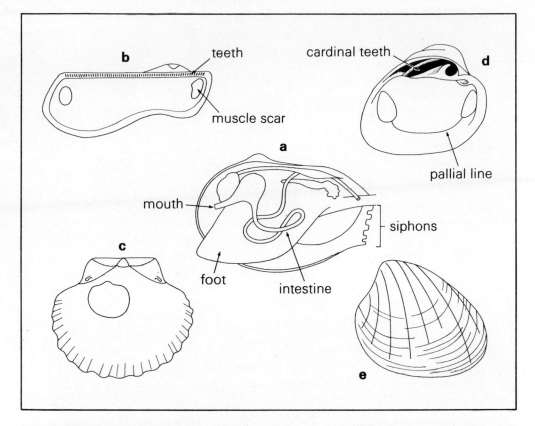

Bivalves. **a** View of bivalve with left valve removed; **b** *Arca*, taxodont dentition, equal muscle scars; **c** *Pecten* dysodont dentition, single muscle scar; **d** *Venericor* heterodont dentition, unequal muscle scars; **e** *Venericor* external view of valve to show growth lines and longitudinal ornament.

developed hinge line lacking teeth. The group is called the cryptodont or toothless bivalves and modern day representatives of the order exhibit simple gill structures and lack a distinct pallial line. The mode of occurrence of these forms in Palaeozoic sediments suggests that the majority were burrowers.

Other more specialized bivalves occur at the same time as the cryptodonts. At an early stage in their development these bivalves adapted to three main modes of life. Some like the cryptodonts adapted to a burrowing mode, whilst others chose fixed or free-living lifestyles. The mode of life of the animal is reflected in the shape and morphology of its shell and an increase in the variety of bivalves throughout the Palaeozoic reflects successful occupation of many different habitats.

As in the cryptodonts, the shells of *Nuculites* and its relatives are thin and smooth. Their hinge lines are noted for the presence of numerous subequal teeth and living representatives spend at least part of their lives within the sediment. *Nuculites* is, like the Recent *Nucula*, included in the Palaeotaxodonta. The ovate to sub-triangular shells of this subclass represent some of the dominant shapes found amongst burrowing bivalves.

A dentition similar to that described above is found in the ark shells, the earliest of which are recorded from the Lower Ordovician. Initially their shells are circular in outline, but later forms are trapeze-shaped with a long, straight hinge line. Many adult ark shells rest unattached on top of the substrate whilst others, particularly in the earlier stages of development, are fixed by byssal threads. The Arcoida are thought to be amongst the least specialized groups of the subclass Pteriomorpha.

Prominent amongst the marine bivalves of the Palaeozoic were numbers of essentially toothless pteriomorphs. Scallops and mussels belong to this group and their earliest representatives date back to Ordovician times. Recent scallops (Pectinacea) are active forms, some being able to propel themselves for short distances through the water. The mussels are mainly fixed or nestling organisms.

Most bivalves are equipped with two equal or subequal muscles which by contraction close the shell. One muscle occurs anteriorly, the other posteriorly. This condition is regarded by many as primitive or conservative and the change to unequal musculature, or to a single, large, often central muscle as being progressive.

In forms with unequal or single muscles, it is the posterior muscle that survives and enlarges at the expense of the anterior one. The enlargement of the posterior muscle is often linked with the elongation of the shell posteriorly. A large single muscle is found in both rounded and elongate shells. Bivalves having this type of musculature may be fixed bottom dwellers or active crawlers and swimmers.

In the geological record the pteriomorphs, as noted above, first appear in the Lower Ordovician, becoming very abundant during the Devonian period. Mussel-like bivalves and pectens are represented by numerous genera and species, and have been recorded from many areas of the world. In the Carboniferous other important families join the established pteriomorph stocks. Some forms are almost indistinguishable from the modern *Pinna*, and *Palaeolima* is an obvious ancestor of the free-living *Lima*. Unlike *Lima*, the members of the Pinnidae are burrowers. Recent forms living with the pointed anterior end of the shell buried in soft mud. The shell is fixed to buried stones or shells by hair-like strands of the byssus. Many pectens appear in the Carboniferous which show similar structures to Recent forms. Other pectens become the first bivalves to be cemented to the substrate, and the first true mussels to bore rocks also appear for the first time.

Many Carboniferous genera persist into the Permian, to be joined there by several new pteriomorph families. One genus, *Eurydesma*, a very characteristic form restricted to Permian sediments, illustrates the progress that can be made through detailed collecting. First described from Australia, it has since been discovered in Pakistan, Kashmir, South Africa and South America.

In palaeoheterodontid bivalves the valves are of equal size and the hinge teeth of either the schizodont or taxodont type. Like the pteriomorphs, the group ranges in time from the early Palaeozoic to the present day. The earliest palaeoheterodontids appear in the middle Cambrian and it is thought that the group gave rise to both the Pteriomorpha and the more specialized Heterodonta. The geological record suggests that the earliest forms were a varied group from which a number of non-marine stocks developed with the first, the genus *Archanodon*, appearing in the Devonian. In the Carboniferous, non-marine bivalves are very common, being associated locally with coal-bearing strata. Specific forms like *Anthracosia*, *Carbonicola* and *Anthraconaia* are very important in coalfield geology. Relatives of the Carboniferous bivalves persist into the Permian, with one fauna found in Brazil and Uruguay seemingly isolated in an enclosed sedimentary basin. The earliest rock-boring bivalve *Corallidomus*, a palaeoheterodontid, is recorded from Upper Ordovician rocks.

The non-marine bivalve *Carbonicola* from the Coal Measures, Upper Carboniferous of Europe. 2·6 cm (1 in) long.

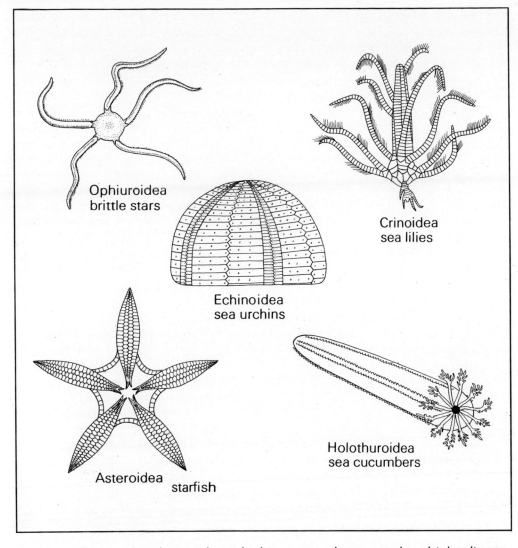

Ophiuroidea
brittle stars

Crinoidea
sea lilies

Echinoidea
sea urchins

Asteroidea
starfish

Holothuroidea
sea cucumbers

The Echinodermata (spiny skinned animals) constitutes a large and varied group of invertebrate animals.

Schizodont palaeoheterodontids have two large teeth which diverge below the beak of the right valve. On the opposite valve there is a central tooth and lateral sockets for the teeth noted above. Bivalves referred to the Schizodonta are divided into two groups. The first are somewhat triangular in outline, many having distinctive surface ornament. These are the Trigoniacea, the earliest representative of which is *Lyrodesma* of Ordovician and Silurian age. The other schizodont stock, the Cardiniacea, are known mainly as brackish and freshwater dwellers, with the form *Amnigenia* occurring in abundance in the Devonian rocks of New York State. By late Palaeozoic times the schizodonts and the Cardiniacea in particular were widely distributed throughout the eastern states of North America and Europe.

Babinka from the lowest Middle Ordovician is one of the earliest bivalves and certainly the earliest heterodontid. Impressions of many pairs of muscle scars are thought by some workers to make *Babinka* transitional between the bivalves and a segmented molluscan ancestor. It is also thought to be ancestral to the lucinoid bivalves which range from the Middle Silurian to the present day. Like the lucinoids, *Babinka* is thought to have been a burrower.

The heterodonts are not much in evidence during the later Palaeozoic times. They are overshadowed by the abundance of non-marine bivalves and genera of the subclass Anomalodesmata.

This last is the final subclass to be considered, the first representatives again appearing in the Middle Ordovician. Most forms have thickened shells and poorly developed cardinal teeth. Many live as shallow nestlers or fissure dwellers. In the Permian, members of the family Megadesmidae reach considerable size in spite of their burrowing mode of life.

In the main, bivalves are generally overshadowed by other invertebrate groups in Palaeozoic stratigraphy, the exception being the non-marine forms

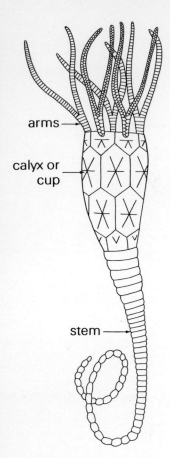

Macrocystella, an eocrinoid echinoderm.

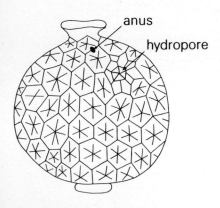

Echinosphaerites, a cystoid echinoderm.

of the Carboniferous. The conservative nature of most clams makes it difficult to recognize distinct evolutionary lines and the group is used mainly in the interpretation of past environments.

The Echinodermata or 'spiny-skinned' animals are highly organized, marine invertebrates. They comprise one of the most characteristic and important phyla of the geological record. The first forms appear in the Lower Cambrian.

Structurally the echinoderms are unlike any other invertebrate group, their origin belonging to the distant eons of the Precambrian. The animal has well-developed nervous, digestive and reproductive systems, and a water-vascular system which consists of an assemblage of canals and folds of the body wall, the tube feet. These have hydraulic and respiratory functions. The body wall consists of three layers, the thick middle layer of several groups being partly calcified. It is these calcified tissues that make the echinoderms important members of Palaeozoic faunas.

In many texts the echinoderms have been subdivided on mode of life, the free-moving forms being called the Eleutherozoa, the fixed varieties the Pelmatozoa. In recent work the phylum has been subdivided into four subphyla and twenty classes. Three subphyla, the Crinozoa, Asterozoa and Echinozoa, possess radial symmetry, whilst the other, the Homalozoa, are assymetrical. The Crinozoa and Echinozoa are present in Lower Cambrian rocks, the Homalozoa appear in the Middle Cambrian and the Asterozoa in the Lower Ordovician.

The enormous variety of echinoderms in the geological record prevents an adequate coverage here, but an attempt is made to show the contribution made by important members of the phylum to Palaeozoic faunas, and to indicate the major trends that took place within various lineages. Of the twenty classes, fifteen are confined to Palaeozoic times and the rest originate during this era.

Of the Crinozoa, the vast majority are attached more or less permanently to the substrate. The symmetry of the group is five-fold or pentamerous. A globose or cup-shaped body is characteristic and various appendages occur which are used for gathering food. The mouth is on the dorsal surface, with the anus situated either dorsally or on the side of the cup. Eight classes of varying importance are referred to the Crinozoa.

The oldest crinozoans, the Eocrinoidea, occur first in the Lower Cambrian. They have sac-like bodies, food-gathering appendages and are fixed by means of a stem. The Cambrian genus *Macrocystella* is typical of the class, having four vertically arranged rows of five plates. Each top plate bears a food gathering appendage, which is divided into two. The plates of the eocrinoid calyx are of solid calcite, and lack perforations, which distinguishes them from more specialized groups. Eocrinoid arms are, however, similar to those of the cystoids and may suggest some ancestral link. The last eocrinoids occur in rocks of middle Ordovician age. Three other crinozoan classes are confined to the middle Ordovician period. At first the crinozoans are of limited importance in Lower Palaeozoic faunas, but with the appearance of the Cystoidea and Crinoidea in the early Ordovician, the representation of the echinoderms in shallow shelf communities increases dramatically.

The cystoids, although more advanced than the eocrinoids are still regarded as primitive, stemmed echinoderms. The group is a varied one and its members occur throughout the Palaeozoic. Most forms are radially symmetrical but a few have bilateral symmetry. Only a few species have a wide geographical distribution, although a single species could thrive locally. Several authors have suggested that cystoids flourished only in areas of abundant food. Large colonies appear to have existed in relatively quiet waters, the genus *Echinosphaerites* occurring in masses in clay-rich areas of the Ordovician. A few cystoids probably became free-living, moving slowly over the bottom muds. The geological record of the cystoids is overshadowed by that of the Crinoidea and Blastoidea.

The latter are characterized by their pronounced symmetry and the regular arrangement of the thecal plates. Geologically the blastoids extend from the

Middle Ordovician to the end of the Permian. Evolutionary changes occur, but do not appear to follow any real direction. Early forms resemble the cystoids, but differences occur in pore structure. Silurian blastoids are noted for their five-fold symmetry and many Devonian genera for the development of a hydrospire. In the Lower Carboniferous the blastoids reach a peak of development and variety. The form *Pentremites* is particularly abundant during this period. From field data it would appear that blastoids were gregarious like other echinoderms. Their size varied with the type of sediment they lived on, large individuals being associated with finer sediments. Symbiotic associations between blastoids and gastropods have been discovered in the fossil record.

The Crinoidea are the dominant class of the Crinozoa, their round or pentagonal stem ossicles being discovered locally in 'drifts' throughout the stratigraphic record. The majority are fixed benthic forms, but secondary, free-swimming forms are known both as fossils and at the present day. The morphology of the crinoids is quite characteristic. The calyx is globular or sac-shaped, like other crinozoans, and the calcareous plates are symmetrically organized. Food-gathering arms occur, and, like the stem, are made up of numerous calcareous ossicles. Anchorage structures are present in many forms. Numerous evolutionary trends occur, affecting the number of plates in the calyx, the arms and other specific features. The calyx of crinoids is divided into a dorsal cup and tegmen (ventral portion of body between arm bases), and the classification of the group is based in part on the structure of the cup.

The crinoids can be divided into four subclasses, the Inadunata, Flexibilia, Camerata and Articulata. Of these, the first three are confined to the Palaeozoic and the last to post-Triassic communities.

The precise origins of the class are lost in the Precambrian, although both cystoids and eocrinoids have been proposed as possible ancestors. Current views on the origin and the evolution of the group are marked by extreme

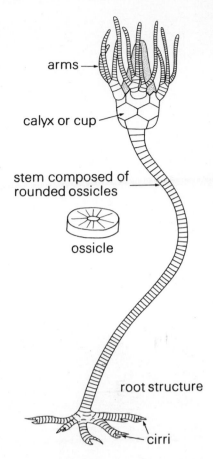

Crinoid structure.

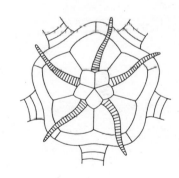

Monocyclic cup.

Dicyclic cup.

Left
Pentremites (Carboniferous) exhibits the five-fold symmetry characteristic of the blastoid echinoderms. 2 cm (0·8 in) in diameter.

caution. One theory suggests that the rigid union of plates in the camerates separates them from the main line of crinoid evolution. Supposedly the camerates and inadunates evolve from a common ancestor, and in turn the inadunates give rise to both the Flexibilia and the Articulata.

Throughout the Palaeozoic, the inadunates and camerates dominate shelf-sea faunas. The early profusion of inadunate genera in the Middle Ordovician was overshadowed in the Silurian, Devonian and Carboniferous by many hundreds of camerate species. Peaks in crinoid numbers and diversity are seen in the reef-building phases of these periods, when even the outnumbered Flexibilia show a considerable increase in abundance. At the end of the Lower Carboniferous the camerates show a dramatic reduction in species from over 700 to 4. This reduction is also seen in the other subclasses, but they persist in significant numbers until the end of the Permian. The first free-swimming stemless crinoid, an inadunate, appears at the end of the Lower Carboniferous.

Of the remaining echinoderms the Echinozoa are the major group, ranging in time from the Lower Cambrian to the present day. Unlike the Crinozoa, they are free-living, cylindrical, globular or disc-shaped animals. They are radially symmetrical and lack outspread food-gathering appendages. Modern echinozoans are algal browsers, scavengers or predators. In primitive forms

Sea lilies (crinoids) were amongst the more common animals found in Silurian sea-floor communities. Length of arms and calyx 10·5 cm (4 ins).

the mouth and anus occur at opposite ends of the body, while in more advanced types the mouth is ventral and the anus either dorsal or remote from the oral area. In active forms the tube feet, normally associated with respiration and feeding, aid locomotion. Appendages (spines or articulated grasping organs) on the plates of some types also assist in locomotion. Of the seven classes referred to the Echinozoa, four are of importance in the geological record. The others are rather poorly represented and somewhat enigmatic. One group, the Ophiocistioidea, were free-moving echinozoans, the body partially covered or fully enclosed by a test of large calcareous plates, the tube feet on the undersurface reaching extraordinary proportions.

Unlike the other classes, the sea cucumbers or Holothuroidea have a spicular skeleton, scattered spicules representing the group from the Lower Carboniferous to Recent times. Apart from the doubtful form *Redoubtia polypodia* from the Burgess shales, they are unknown in the Lower Palaeozoic.

Helicoplacus from the Lower Cambrian of North America, has a flexible and spirally pleated test. Structurally it is similar to the Echinozoa but no definite link has been established. Echinozoans with flexible tests appear in the Lower Cambrian. These are the Edioasteroidea, the general morphology of which suggests a fixed mode of life. The test is made up of numerous small plates, with the mouth occurring centrally on the dorsal surface. Five rows of pored plates (the ambulacra) radiate outwards towards the margin of the test. No stem is present, some forms being directly attached to hard substrates, others to shells or rock fragments. A few may have burrowed, but most lived in shallow, littoral environments. Association with other echinozoan stocks are doubtful, but the structure of the rows of pored plates is comparable with those of the Echinoidea or sea urchins, the earliest representatives of which appear in the Ordovician. The class has a rather box-like internal skeleton made up of many plates, and the mouth is placed on the ventral surface. The main area of the skeleton, the corona, is divided into rows of plates. Spines may occur over the surface of the test.

Lecythocrinus eifelianus, a Devonian crinoid, pictured here with a brittle star (Eifel, Germany). Length of brittle star from arm tip to arm tip 4·5 cm (1·8 ins).

The echinoid *Hemicidaris*.
a Side view; **b** ventral view
(mouth surface); **c** dorsal view
(anal surface). In life, the
spines articulate with the
surface of the mammelons.

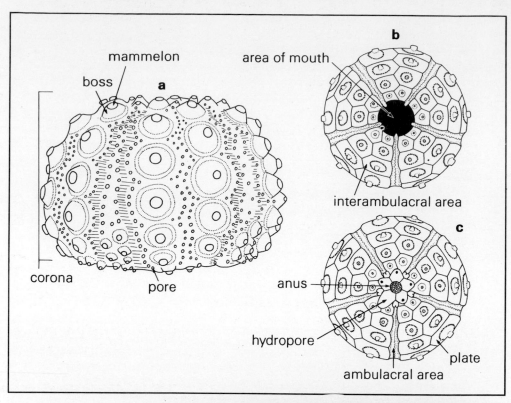

The sea urchin, *Archaeocidaris*,
Lower Carboniferous, Campsie,
Scotland. Width of matrix
11·8 cm (4·6 ins).

Flexible and rigid tests are found amongst Ordovician echinoids. The flexible variety, however, are thought to be the more primitive and, by many workers, to be ancestral to later stocks. The first echinoids are rounded but throughout the Palaeozoic a general trend leads to flattened varieties by the dawn of the Lower Carboniferous. Echinoids with flexible tests dominate the history of the group throughout the Ordovician and Silurian. Certain lineages show a reduction in number of plates and plate rows. In the Devonian period members of the Cidaroida appear which have only two rows of simple plates in each ambulacra. Early cidaroids had flexible tests and more than two rows of plates in each interambulacra. By the end of the Palaeozoic many cidaroids had rigid tests, with two rows of plates in each plate area. Unlike their ancestors, the cidaroids lived in shallow waters. The Cidaroida gave rise to all the echinoid stocks of post-Palaeozoic times.

Rhodocrinus, a crinoid from the Lower Carboniferous of Iowa, USA. 3·5 cm (1·4 ins) wide.

The other two echinoderm classes, the Asterozoa and Homalozoa, also have their origins in the Palaeozoic. The Homalozoa, the 'flat-animals', are exclusively Palaeozoic, ranging from the Middle Cambrian to the Middle Devonian. The Asterozoa occur somewhat later, ranging from the Lower Ordovician to the present day. Both groups occur in rather limited numbers in the earlier part of the fossil record, but both are important from an evolutionary point of view.

Within the Homalozoa a number of orders occur which were previously placed in the 'Carpoidea' of other classifications. The latter were regarded as primitive fixed echinoderms, although note was made that they resembled some primitive chordates. Unlike other echinoderms, the homalozoans are asymmetrical, with many forms having a limited number of very large plates on one side. Detailed analysis and dissection of fossils belonging to this group has revealed an enormous amount about soft-part morphology. Some workers have, on the basis of painstaking reconstruction, suggested that certain carpoids possessed gills and highly organized vascular and nervous systems. Bravely, in the face of academic criticism, these workers have proposed that certain homalozoan stocks are linked with chordate evolution. Current ideas suggest that the majority of homalozoans rested on or burrowed into the substrate. In this, they bear comparison with their asterozoan cousins.

The starfish and brittlestars (Asteroidea and Ophiuroidea) are free-living bottom dwellers. Some are active burrowers, whilst others move freely over the substrate in search of food. Asterozoans have a five-fold symmetry with the flattened body composed of a central disc and five arms. The mouth is usually ventral and five ambulacra radiate outwards to the tips of the arms. Each ambulacra is crowded with tube feet, which are coordinated by a well-developed nervous system, enabling efficient crawling or burrowing. Starfish were, without doubt, important members of the vigorous animal communities that existed during the Palaeozoic. Usually, however, their remains are preserved as separated ossicles, although occasionally 'starfish beds' are discovered.

Many of the phyla considered in this account of the Palaeozoic have been bottom dwellers, and the majority are far removed from chordate evolution. The homalozoan echinoderms have, as suggested, been linked with chordate development; but this link is questioned by many palaeobiologists. Another link, not recognized until recent times, does however exist. The class Grapto-

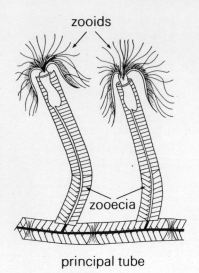

zooids

principal tube

zooecia

Rhabdopleura, a Recent relative of the graptolites.

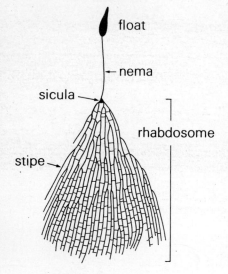

float

nema

sicula

rhabdosome

stipe

The dendroid graptolite *Dictyonema*.

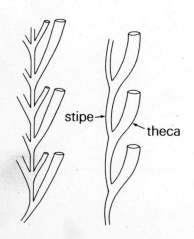

stipe

theca

The growth patterns of **left** a dendroid graptolite and **right** a monograptid.

The graptolite *Monograptus* is characterized by the presence of the cup-like thecae on only one side of the stipe. 4·5 cm (1·8 ins) long.

The whole graptolite colony, be it dendroid or graptoloid starts as a single bud, subsequent buds developing to form branches. 1 mm (0·04 ins) long.

lithina of the Palaeozoic is referred to the phylum Hemichordata. Graptolites and a host of ancient relatives are ranked alongside hemichordates, such as *Rhabdopleura*, through the basic similarity of their chitinous skeletons. The soft parts of the graptolites are unknown but, like the living rhabdopleurids they were colonial and it is thought that minute plumed animals lived inside tube-like structures (thecae). The graptolites are divided into two orders, the Dendroidea and Graptoloidea.

In the former, colonies may have hundreds of branches which bear thousands of individual thecal structures. The vast majority of dendroids were fixed to the sea-floor growing as shrub-like or inverted cane type colonies. In life it is probable that some individuals carried on feeding and reproductive functions, whilst others kept the colony free of sediment deposition. Some dendroid genera adapted to a free-swimming or attached planktonic life, whilst others such as the Anisograptidae undertook a reduction in the number of branches and thecal variety. The anisograptids were ancestral to the graptoloids.

Graptoloids continue the trend in branch reduction, the ultimate form being the single-striped monograptids (Silurian). Many graptoloids are both short-lived and widely distributed and as a result they make ideal zone fossils in the Ordovician and Silurian Periods. The thecae of some graptoloids become comparatively large and isolated, reflecting an increase in size of the occupant. Others become rather elaborate structures, suggesting the need for protection against a foe. Evidence has been put forward to show that the individual animals moved in a coordinated manner, currents created during movement assisting feeding and respiratory functions.

The abundance of graptoloid remains in black shales suggests that the majority were planktonic or free-swimming. Morphological evidence to support this hypothesis includes the discovery of float structures and the microstructure of the skeleton. The last graptoloids are known from Devonian rocks, and the last of the conservative dendroids from Carboniferous rocks.

All the groups described so far are invertebrates, their membership of Palaeozoic communities remaining unchallenged until early Devonian times.

Phyllograptus elongatus, a 'leaf-like' graptolite from the Lower Ordovician, Norway. 1·8 cm (0·7 ins) in length.

Graptoloids have fewer branches than dendroids, *Dichograptus octobrachiatus* from the Ordovician having eight (specimen from Australia). Width of colony 12·1 cm (4·7 ins).

Single-stiped graptolites mark the final stage in the reduction of the number of branches. *Orthograptus rugosus* (Upper Ordovician, Laggan Burn, Scotland) is a fine example of them. Width of colony 5·6 cm (2·2 ins).

A two-stiped graptolite *Didymograptus murchisoni* (Ordovician, Shropshire, England). Length of stipe 3·2 cm (1·2 ins).

In the Lower Palaeozoic they cohabited with the algae the majority of shallow-water environments. Seaweed-like fossils, including the genus *Newlandia*, are found in Cambrian and Ordovician marine sediments.

The earliest vertebrate remains are known from the Upper Cambrian of Spitzbergen, Norway, and Ordovician sandstones of North America and Russia. These fragments probably belong to jawless fish (Agnatha) called the ostracoderms, fossils of which are well-known from Silurian and Devonian rocks. These are a primitive stock, and fish with jaws, the placoderms, represent the true root stock of fish evolution, for from these the cartilaginous (sharks) and bony fish evolved by the Middle Devonian. The abundance of ostracoderms and placoderms in Devonian rocks lacking invertebrate faunas makes them useful stratigraphic tools.

Through a number of structural and physiological changes, tetrapods in the form of the Amphibia arose from the lobe-finned bony fish in late Devonian times. Like the ostracoderms, the earliest amphibians were restricted to the Northern Hemisphere, and have only been recorded from the continents of Europe and North America. This restricted geographical distribution persisted throughout the Carboniferous period. Whilst the Amphibia, and later the Reptilia, set about the conquest of the terrestrial environment, fish became the dominant predator and scavenger group of the Late Palaeozoic. Their abundance and diversity has been used, by many workers, to explain the demise of the trilobites and graptolites. In part this may be true, but the diversity of fish also diminishes temporarily in Late Palaeozoic times.

Throughout the Palaeozoic, peaks, troughs and plateaus in the number of new groups, occur in the fossil record. Major peaks are recorded during the Cambrian-Ordovician, Devonian and late Permian. Significant extinctions occurred in the Middle Ordovician, Devonian and at the end of the Permian. The appearance and disappearance of skeletonized shelf-sea invertebrates have been linked with environmental factors. Periods of steady conditions would not encourage the development of new species, whereas environmental fluctuations cause greater localization and diversification. The extinction of the trilobites, rugose corals, certain brachiopods, eurypterids and numerous echinoderm, foraminiferid and bryozoan families cannot be attributed to any single factor. The extinction of certain land plants and tetrapod stocks probably required different conditions to those of the marine invertebrates,

Bothriolepis, Devonian, arthrodiran fish. During the Devonian period many areas of the world were subject to drought, lakes dried up and fish faunas were left to die on the muddy substrate. Head shield 10–15 cm (4–6 ins) long.

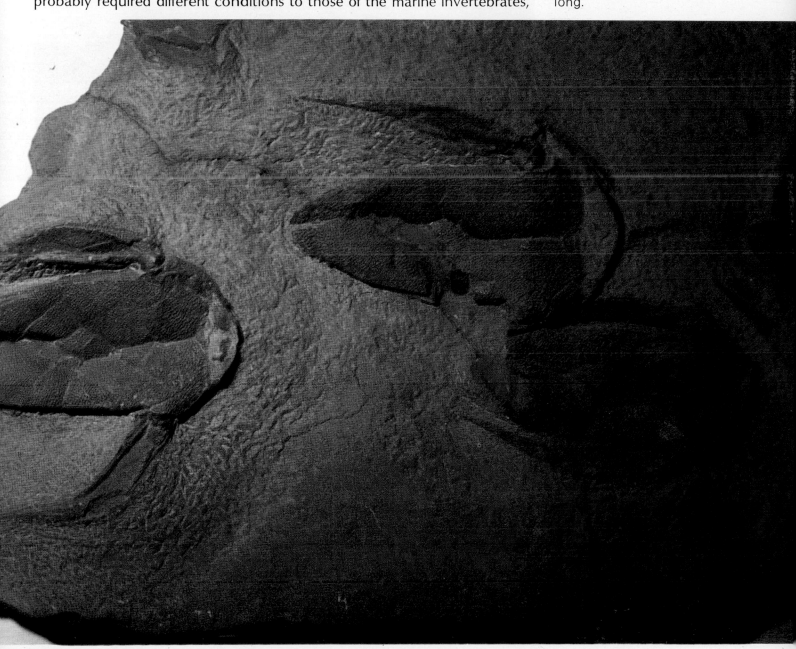

Detached leaves of the spore-bearing horsetail *Annularia*, Coal Measures, England. Rock sample 11 cm (4·3 ins) wide.

The head shield of the Devonian fish *Cephalaspis*. 9 cm (4 ins) wide.

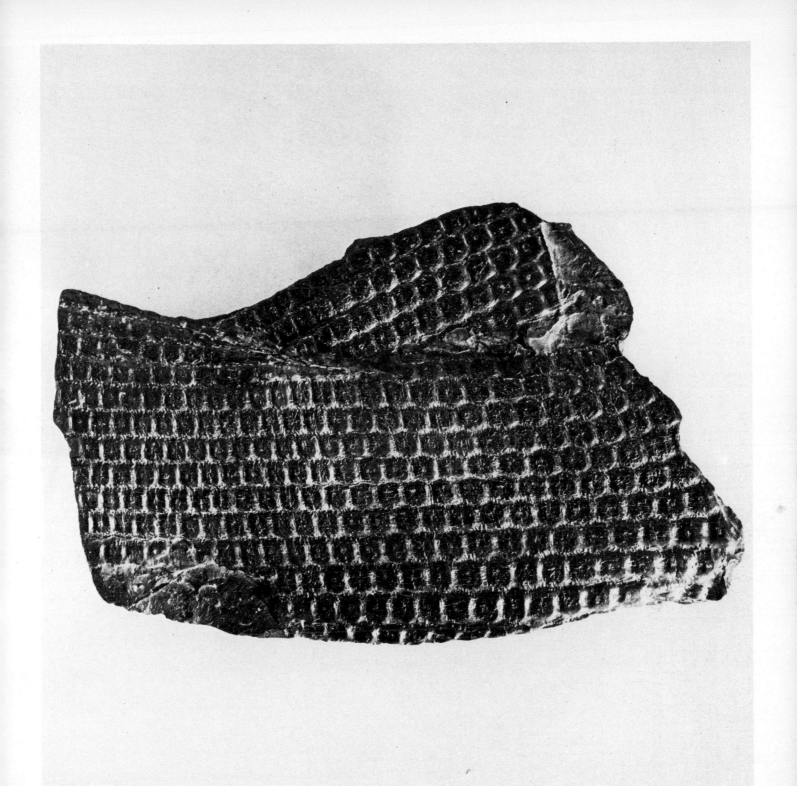

but overall climatic fluctuations may have resulted in the same overall effect. Mountain-building episodes may have emphasized these climatic changes and certainly caused the withdrawal of seas from areas hitherto populated by shelf-sea communities. The draining of coal-measure swamps affected the Amphibia and swamp-dwelling plants.

The extinctions of Palaeozoic faunas and floras may at first appear to have been the result of a sudden catastrophe, a theory supported by many a palaeontologist, past and present. They are, however, more likely to be the result of slow change. The persistence of local faunas such as the Permian blastoid and crinoid communities of Timor indicate that some areas offered stable enrivonments which protected some species from extinction, for elsewhere in the world blastoids are unknown in rocks younger than the Carboniferous.

During the Carboniferous, great coal swamps covered much of the Earth. Numerous plant species, such as *Sigillaria* and *Lepidodendron* characterized the faunas of this age. *Sigillaria elegans*, a Coal-Measure plant from the Carboniferous of Liege, Belgium; stem covered with leaf scars. 20·2 cm (7·9 ins) long.

The Mesozoic era

The dawn of the Mesozoic era is noted for the spread of marine areas during the early Triassic. With these came a veritable explosion of new marine stocks, many replacing forms that had died out during the Permian. Replacements and takeovers occur throughout the fossil record, but those of the Palaeozoic-Mesozoic boundary are amongst the more spectacular. Eighteen or so orders disappeared during the Permian, leaving behind a large number of empty ecological niches. These were occupied by new forms during the Triassic and Jurassic, with the appearance of nearly 200 new families.

An example of ecological replacement can be found in the study of brachiopod and bivalve stocks. The demise of the productid brachiopods, which themselves replaced the pentamerids as soft-mud dwellers, left a vacant niche into which the thick shelled bivalve *Gryphaea* evolved. The general reduction in brachiopod stocks coincides with the rise of more efficient bivalves such as the oysters.

To invertebrate palaeontologists the Mesozoic is the era of the mollusc, whilst to vertebrate workers the period is known as 'the age of the reptiles'. Both groups are undoubtedly very important, but other stocks also make major contributions to Mesozoic history. As in Palaeozoic times, the fossil record of the Triassic, Jurassic and Cretaceous periods exhibits strong fluctuations. Numerous ammonoids and brachiopods become extinct at the end of the Triassic, a floral change takes place during the early Cretaceous and a number of extinctions occur at the end of the Mesozoic. These are accompanied or followed by the appearances of many new families in the Jurassic and Tertiary periods. The changes that took place in specific phyla and their respective roles in communities of the Mesozoic era are set out in more detail below.

Of the protozoan families that dominated during the Palaeozoic, the fusulinids have no representation in Triassic rocks. In fact, few microfossils have been found in sediments of this period. Not until the Jurassic do the foraminiferids regain some of their former importance, with the families Lagenidae and Polymorphinidae being prominent. Most genera referred to these families have calcareous tests and are bottom dwellers. In the Cretaceous, microscopic foraminiferids contribute greatly to the formation of the chalk deposits of northern Europe and the Gulf Coast of America. Planktonic foraminiferids flourished during the Upper Cretaceous, with many genera, related to the Recent form *Globigerina*, floating in the warm waters. The so-called larger Foraminifera also appear in the Lower Cretaceous, with agglutinated forms occurring first. Large calcareous foraminiferids abound in the Upper Cretaceous, with over thirty new families arising throughout the world. The larger foraminiferids and planktonic globigerinids are extremely valuable index fossils during the Cretaceous. They also assist in palaeogeographical reconstruction. Radiolarian oozes are also abundant during the Mesozoic era.

Gryphaea incurva flourished on soft muddy sediments during Jurassic times. 7 cm (2·7 ins) high.

Siphonia koenigi, a glass sponge from the Upper Cretaceous of Europe. 23 cm (9 ins) long.

Calcareous and siliceous sponges flourish during the Mesozoic. Siliceous forms are indicative of rather cold, moderately deep waters and localized communities occur throughout the Mesozoic. Those of the chalk of Yorkshire and northern France yield beautiful examples of the glass sponges. Isolated glass sponges (Hyalospongea) such as *Ventriculites* are common in Upper Cretaceous sediments.

Calcareous sponges often differ from siliceous types in having thicker walls and rather irregular shapes. They are found in shallower, warmer waters, the Farringdon sponge gravels of the Lower Cretaceous of Berkshire, England, being deposited close to the shoreline of the ancient London platform. Over seventy species of sponge have been recorded from these deposits. Small brachiopods, bryozoans and oysters are associated with this fauna.

After the disappearance of the rugose and the majority of the tabulate corals at the end of the Permian, an hiatus in coral evolution took place during the Lower Triassic. No intermediate forms exist between the Permian and Middle Triassic, although a few genera, confined in time and distribution to the upper part of the Lower Carboniferous in Scotland and Germany, do show certain affinities to the Middle Triassic Scleractinia (hexacorals). Like their probable ancestors, the scleractine corals are polypoid coelenterates, the polyp being almost identical to that of a sea anemone. Unlike the Rugosa, the scleractine skeleton has septa inserted in multiples of six.

Middle Triassic hexacorals are limited in numbers and diversity. Upper Triassic forms are relatively abundant, with both solitary and colonial corals contributing to reef growth. The distribution of Upper Triassic hexacorals is along a rather narrow equatorial belt. The composition of early Jurassic coral communities is similar to that of late Triassic times. In the middle and late Jurassic, however, a considerable expansion took place, with scleractine corals and reefs being abundant in the shallow shelf seas of the period. *Montlivaltia* and its relatives are common throughout the Triassic, late Jurassic and early Cretaceous. *Thamnastria* and *Isastrea* form reef masses in the Jurassic of the British Isles.

After a period of abundance in the Lower Cretaceous, the scleractines undergo a temporary decline until the next period of reef building in the Upper Cretaceous.

Unlike the rugose polyps, those of the hexacorals overlap the outer wall of the skeleton and according to various workers this allows for modifications of the outer edge of the coral cup and the greater development of skeletal tissue between polyps, two changes that apparently gave impetus to an increase in colonial development amongst Mesozoic scleractines.

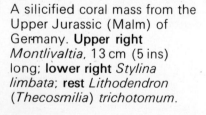

A silicified coral mass from the Upper Jurassic (Malm) of Germany. **Upper right** *Montlivaltia*, 13 cm (5 ins) long; **lower right** *Stylina limbata*; **rest** *Lithodendron* (*Thecosmilia*) *trichotomum*.

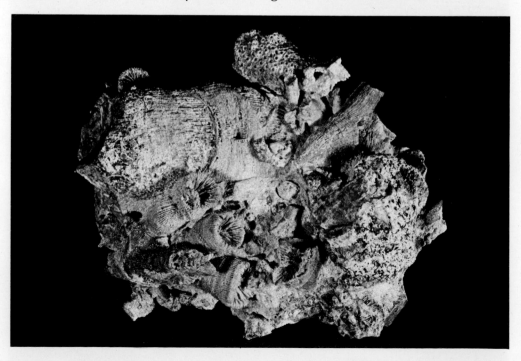

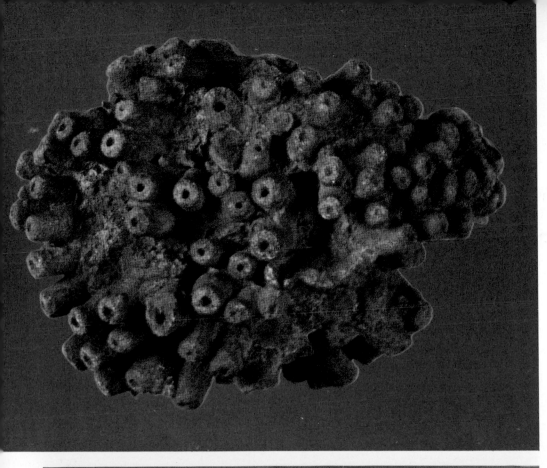

Calcareous sponges such as *Peronidella* flourished in the Farringdon region of Berkshire, England during Lower Cretaceous times. Individual unit 7·5 mm (0·3 ins) in diameter.

Entobia cretacea was a boring sponge. Casts of its activities are often found preserved in Cretaceous flint nodules. Single burrow cast 2 mm (0·08 ins) long.

Aspidiscus cristatus, a hexacoral from the Cretaceous of Chetaibi, Constantine, Algeria. 5·7 cm (2·2 ins) across.

Apart from the Scleractinia other polypoid coelenterates, the sea pens and sea feathers (Pennatulacea) occur in the Mesozoic. Their skeletons are either horny or calcareous, the latter being often of a spicular nature. Of the limited number of genera that existed in the Mesozoic, *Heliopora* survives as a modern reef builder.

The number of medusoid coelenterates preserved in sediments of Mesozoic age are probably no greater than those of the Palaeozoic, although a significant number of species have been recorded from the Solnhofen limestone (Jurassic) of southern Germany. *Kirklandia texana* of the Cretaceous is similar to the form *Brooksella* from the Middle Cambrian.

Few, if any, of the Palaeozoic bryozoans survive into the Triassic, and no apparent link exists at this time between the ancient families of the Palaeozoic and the important orders of the Mesozoic and Cainozoic eras. Obviously some link must exist between the cyclostome bryozoans, common to all eras, but the association of the Cheilostomata and the extinct Palaeozoic groups is much more tenuous.

In the Jurassic the cyclostomes reappear in some numbers, with forms like *Entalopora* and *Spiropora* flourishing. Cheilostome bryozoans appear for the first time in the Jurassic and, like the cyclostomes, undergo a major expansion during Cretaceous times. Unlike other bryozoans, certain cheilostomes encase the zooid in a calcareous skeleton, the animal emerging to feed through a small operculum.

Cheilostome colonies, although rather small and delicate, are often complex and extremely beautiful. The complexity and short stratigraphic range of certain species makes them important index fossils.

As with the bryozoans, many brachiopod orders die with the Palaeozoic era. The orthids and productids may have survived into the Triassic but in general they, like other groups, failed to overcome the suspected ecological traumas of the late Permian. Spiriferids persisted into the Middle Jurassic, but essentially the Mesozoic is known for the abundance and diversity of the Terebratulida and Rhynchonellida.

It has been suggested that these two orders, together with various inarticulates, survived the cold glacial spell and the expansion of terrestrial areas in the Upper Carboniferous and Permian because they were more capable of change, being used to a shallow-water habitat.

The rhynchonellids undergo a number of evolutionary bursts in the Mesozoic, with different stocks dominating each of the three periods. A detailed analysis of the various habitats occupied by brachiopods in the Mesozoic, reveals that the rhynchonellids dominated brachiopod assemblages during the Triassic and early Jurassic, their positions being taken in later Jurassic and Cretaceous times by the terebratulids. Replacement was not restricted to orders, as individual species tended to be replaced in successive horizons. Seven main habitats have been recognized for brachiopods during the Mesozoic Era. These range from rocky coastlines to the depths of the oceans. It is remarkable that the unspecialized inarticulate *Lingula* survived unchallenged throughout, occupying the littoral zone.

Apart from the seven main habitats, a number of provinces can be recognized within the Mesozoic faunas of Europe, Greenland and North Africa. Some forms such as *Flabellothyris*, common to India, Morocco and Mexico, appear to represent climatic zonation; whilst the distribution of certain Alpine faunas is controlled by the type of substrate.

Mesozoic brachiopods are often found in great accumulations on bedding planes, or in 'nests' within sediments; the Middle Jurassic faunas of the Normandy coastline, characterized by the abundance of *Digonella*, and the Lower Cretaceous Greensand faunas of the Isle of Wight, in which *Sellithyris* occurs in closely packed masses, being examples. *Digonella* was an inhabitant of the muddy subtidal zone, whilst *Sellithyris* lived in littoral and sandy subtidal areas.

In the Cretaceous, the abundance of brachiopods in Europe is not paralleled in North America. The sediments and general content of brachiopod faunas are similar but, for some unknown reason, numbers and diversity are very restricted. In certain areas of the world, and in Denmark in particular, Danian sediments link the Upper Cretaceous and Cainozoic. The brachiopods, echinoids and bryozoans of the Danian bridge the gap between Cretaceous and Cainozoic cousins.

Jurassic oysters used this rhynchonellid as a base for attachment. Note the change in shell ornament of the oysters related to the ribbing of the brachiopod. 2 cm (0·8 ins) long.

A 'nest' of Cretaceous brachiopods *Sellithyris sella*, Lower Greensand, Isle of Wight, England. Individual brachiopod 2·4 cm (0·9 ins) wide.

It is possible that the extinctions of the Cretaceous were of less importance, amongst certain groups, than they first appear. Brachiopod stocks suffer more during the Rhaetic and Lower Jurassic, when the deposition of black shales and possible pollution are associated with the decimation of various brachiopod and ammonite stocks.

Bivalves and ammonoids flourish during the Mesozoic era, the latter being used as zone fossils during all three periods. The bivalves had shown considerable diversity during the Palaeozoic and in the Mesozoic replaced the brachiopods as the major group in shallow-water benthic communities. The ammonoids, however, had a very different history, for of the thirty-seven genera present in the Permian, only one actually survives into the Triassic period. Another eight genera are found in the earliest Triassic, several being directly related to two Permian families.

Of the other molluscan groups, the nautiloids continue into the Triassic without any apparent evolutionary change, the ancestors of the first Mesozoic genera being recognized amongst well-known Upper Palaeozoic forms. Like the ammonoids and bivalves, the nautiloids undergo a significant radiation in the Triassic period. New gastropod stocks also appear at this time, the most significant appearance being that of the Mesogastropoda. The amphigastropods finally disappear in the early Triassic.

The bivalve story in early Mesozoic times centres on the various families of the Pteriomorpha and Palaeoheterodonta. The pectens and oysters became well established during the Triassic, with individual genera occurring in great numbers in certain deposits. Amongst the oysters, small, smooth ancestors of the Jurassic genus *Gryphaea* appear in the Upper Triassic of Sicily, whilst the dentate form *Lopha* appears at the same time in the Alps. Of the palaeoheterodonts, the descendants of the Palaeozoic genera *Schizodus* and *Lyrodesma*, the myophorids, reach the peak of their development. From these the first representatives of the well-known genus *Trigonia* arise in middle Triassic times.

Heterodonts are not really common in Triassic rocks but the resurgence of the pachydont *Megalodon* is an important development in bivalve history.

The early Mesozoic bivalve faunas appear to be dominated by widespread and long-ranging genera. Jurassic communities are characterized by the abundance of oysters, trigonids and limas in the pteriomorphs, and the burrowing and boring pholadomyids of the Anomalodesmata. Pectens continue to thrive and many species have a modern aspect.

Gryphaea, a probable replacement for the productid brachiopods, abounds in the Lower Jurassic. Initially it was fixed to the substrate, but in later life just lay in the mud, its dorsal valve acting as a lid. Unlike *Gryphaea*, many oysters are fixed throughout their lives, and it is this that has made them such a successful group. Forms such as *Liostrea* of the lowest Jurassic, cement themselves to pebbles, whilst *Exogyra* and *Ostrea* of the middle and late Jurassic occur cemented to the substrate or crowded together in great masses.

In the Upper Jurassic and Cretaceous, the widespread faunas of early times are replaced by more localized forms. The separation of the continents and major climatic changes are instrumental in the appearance of these geographically restricted groups. Certain genera remain cosmopolitan throughout time, but others are restricted to climatic zones. Of the forms mentioned in this text, *Nucula*, *Lima*, *Gryphaea*, *Lopha* and *Pholadomya* represent cosmopolitan genera, whilst *Arca* is thought to be a warm-water form.

Examples of other bivalves thought to have inhabited mid-temperate and tropical regions are *Venericardia* and *Lithophaga*. Tropical and subtropical communities of Mesozoic times are associated with the ancient sea called Tethys. This extended from the Americas, through the Mediterranean into the Pacific and southern Asia, its coastlines varying with time.

Of particular importance amongst the Cretaceous Tethys communities are the rudistid bivalves which developed in the last Jurassic from pachydont stocks, such as *Megalodon*. Early rudists are only slightly inequivalve but later forms are strongly so, with the two valves differing in most aspects of morphology. Most rudists are cemented to the substrate, and many are

Trigonia costata, Jurassic, Abingdon, England. Shells and moulds of this bivalve are common in certain Jurassic deposits. The moulds are popularly known as 'horse heads'. 7 cm (2·7 ins) long.

Inoceramus concentricus, a member of a bivalve family which was particularly abundant in Cretaceous seas. Folkestone, England. 4·2 cm (1·6 ins) long.

gregarious and involved in reef building. Rudists occur in many African, Asian and Mediterranean countries and are ideal palaeogeographic indicators. Three basic types of rudist are known in which the attached valve is either coiled, and the smaller of the two, conical, or coiled and the larger of the two. The largest of the rudists are found amongst the conical types, with *Titanosarcolites* reaching 2 metres (6·5 feet) in height. Coiled rudists rarely exceed 30 centimetres (12 inches). The walls of the rudists are often very thick (*Durania* has walls 10 centimetres, 4 inches, thick) and superficially the conical forms resemble horn corals with lids.

No rudist is recorded from rocks younger than those of late Cretaceous age. As with other groups that suddenly disappear from the geological record, no real solution can be put forward for the demise of the rudists. Overspecialization, climate and competition with the scleractine corals have been suggested but no research worker is certain on this, and so it remains one of many great palaeontological mysteries.

The ammonoid radiation of the late Lower Triassic is recorded in a dramatic increase of genera, from the original nine to a total of 136. As with the bivalves, a large number of genera are confined to the Tethys region. By late Triassic times over 400 genera of ammonoids had evolved, including some rather strange heteromorphs (varied structural types). The helicoil and straight shells of numerous forms found in shallow shelf sediments represent stocks that were restricted to small basins within the Tethys region.

Ammonoids with ceratitic sutures dominate faunas of Lower and Middle Triassic. The goniatites have almost disappeared, whilst the ammonites become more common, gradually replacing the ceratites as the dominant group by the Upper Triassic. *Ceratites* is an example of a Middle Triassic ceratite,

Ceratites nodosus exhibiting a ceratitic suture line; Triassic, Braunschweig, Germany. 7·8 cm (3 ins) in diameter.

Far right
Arietites bucklandi, an ammonite from the Lower Lias, Lower Jurassic, Keynsham, England. 32·5 cm (12·7 ins) in diameter.

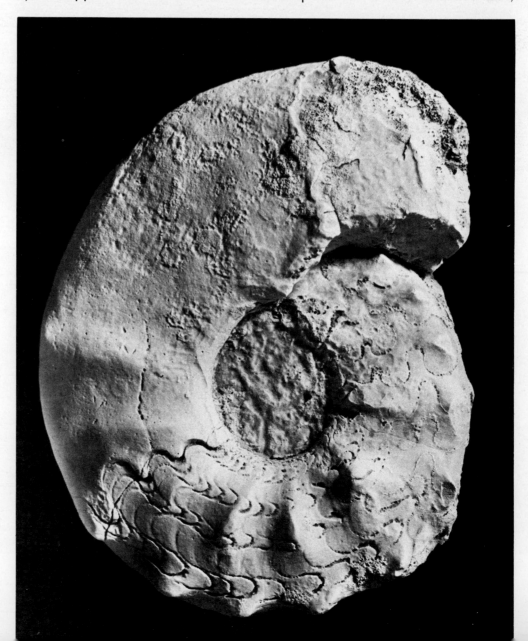

Arcestes (*Proarcestes?*) exhibiting ammonitic suture lines; Upper Triassic, Hallstatt, Austria. 5·7 cm (2·2 ins) in diameter.

whilst *Arcestes* from the Upper Triassic exhibits a suture line of the complex ammonitic type.

By the Jurassic, both ceratitic and goniatitic sutures have disappeared. Jurassic ammonoids developed from several conservative Triassic genera, the coiled and conispiral heteromorphs having vanished.

Three orders of ammonites are recognized, the Phylloceratina, Lytoceratina and Ammonitina and families of all three are present in deposits of early Jurassic age. At first a number of families included in the Phylloceratina and Lytoceratina occur mainly in the Tethyan region, whilst those of the Ammonitina are more widespread. In later Jurassic times palaeontologists recognize a northern (Boreal) as well as a Tethys region. The three orders are well represented in both, although climate and depth may be important factors in the geographic limitation of certain families. Migration routes existed for the dispersal of ammonites during the Jurassic, with Tethys being an important dispersal route for ammonites common to northern Europe, India and East Africa.

The value of the ammonites in Jurassic stratigraphy is expressed in the establishment of fifty-eight zones, which permit an accurate age determination of strata throughout the period.

In the mid-Lower Jurassic, several important families arise, including the Liparoceratidae (*Liparoceras*), Dactylioceratidae (*Dactylioceras*) and the Hildoceratidae (*Hildoceras*). All three are short-lived stocks and, therefore, are of considerable stratigraphic importance. *Liparoceras* and its relatives are essentially northern European, whilst the Dactylioceratidae, important in the Tethys area, are represented by only a few species in areas such as the north of England.

Many ammonites are now thought to be substrate dwellers, and numerous mid-Jurassic heteromorphs are associated with this mode of life. The loss of bilateral symmetry is characteristic of many benthic animals and *Spiroceras* is a typical example. It is probable that the variety of shell form exhibited by the heteromorphs represents adaptation to different modes of life.

Hildoceras is found throughout Europe, Africa and Asia in the lower Jurassic. 4·6 cm (1·8 ins) in diameter.

Ammonite, *Promicroceras marstonense* from the Lower Lias, Lower Jurassic, Marston Magna, England. Individual shell 2 cm (0·8 ins) in diameter.

A detailed study of Upper Jurassic ammonite faunas reveals that the areas noted above are characterized by the appearance and relative abundance of different families, with the arctic Boreal province dominated by one family, the Cardioceratidae (*Cardioceras*). In the Tethys region the phylloceratids are very abundant, but the migration of northern Boreal faunas southwards is regarded as a major feature of ammonite history during this period of time.

Whilst the heteromorph faunas of the Triassic and Jurassic were rather sluggish in their evolution and limited geographically, Cretaceous heteromorphs evolved quite rapidly and were soon represented in all seas. Forms such as *Hamulina*, *Hamites* and *Ancyloceras* represent the uncoiled heteromorphs, whilst *Turrilites* and *Nipponites* are conispiral and coiled asymmetrical types. Not all Cretaceous ammonites are heteromorphic and benthic; many are typically bisymmetrical with involute or evolute shells that suggest a free-swimming or planktonic mode of life.

Turrilites costatus, a spirally coiled ammonite from the Chalk, Upper Cretaceous. 7 cm (2·7 ins) high.

A coiled nautiloid from the Jurassic. 11 cm (4·3 ins) in diameter.

Scaphites nodosus, a Cretaceous ammonite from South Dakota, USA. 4·2 cm (1·6 ins) in diameter.

81

During the Cretaceous, the nature of the suture line of several ammonites is simplified, the sutures of *Tissotia*, for example, being similar to those of the ceratites. Others, like *Baculites*, have straight shells, which may suggest a return to the sea-bottom, scavenging niche previously occupied by the orthocone nautiloids.

At the end of the Cretaceous the ammonoids became extinct. Their absence from rocks of Eocene age has been attributed to many causes, including a sharp increase in the number of sharks or the appearance of aquatic mammals. Mass extinctions appear commonplace at the end of the Cretaceous and it is possible that a common thread exists that could indicate breaks in existing food chains or excesses in predator-prey relationships.

During the Mesozoic the nautiloids are of limited importance, the majority of forms being coiled with simple sutures. The restricted role of the Mesozoic nautiloids is in contrast with the rise of the belemnoids. The massive 'bullet-like' skeletons of these are found in deposits from the Carboniferous to the Eocene. Pre-Triassic and post-Mesozoic discoveries are rare, and the important Belemnitida are associated solely with Jurassic and Cretaceous sediments.

The skeletons of belemnoids are rarely found whole, and the discovery of soft parts is even rarer, although a number of complete specimens have been collected from the Solnhofen limestone (Lithographic limestone, Bavaria, Upper Jurassic). Belemnoid hard parts can be divided into two main units, the chambered phragmacone and the tapering, subcylindrical rostrum. The phragmacone is cone-shaped and partitioned by simple, gently concave septa. The siphuncle is sited at the ventral margin. A large blade-like extension, the pro-ostracum, extends well beyond the last septum. The delicate nature of the phragmacone and pro-ostracum usually means that they are crushed during compaction, whilst the more massive rostrum or guard built up of successive calcite layers survives. In many deposits the guards show a preferred orientation or alignment and can be used to plot current direction. The phragmacone slots into a deep cavity at the near end of the rostrum.

The shell of belemnoids was internal, the pro-ostracum protecting the soft parts and providing a surface for muscle attachment. As in the nautiloids and ammonoids, the chambered phragmacone had a hydrostatic function, whilst the rostrum acted as a balance structure. The soft parts of the belemnoids, like those of their modern relatives the squids, included arms which were equipped with small hooks, used to grasp and hold soft-bodied prey.

The belemnites, are of Jurassic-Cretaceous age and some families achieved a worldwide distribution during these periods. The majority of Jurassic families, however, are best represented in the Northern Hemisphere, and it would appear that the centres of radiation of the group were based there as well. Lower Jurassic forms were confined to the Tethys area. Belemnites are thought to have been inhabitants of shallow shelf seas, and it is possible that the spread of various families in the Middle and Upper Jurassic was related to the establishment of shelf-sea connections between the various provinces. *Megateuthis*, first found in Lower Jurassic deposits, is thought to be important in the evolu-

The skeleton of a Jurassic belemnite. 26 cm (10 ins) long.

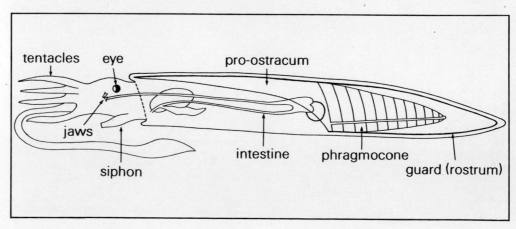

Reconstruction of a belemnoid.

82

tion of later stocks. In post-Lower Jurassic times, belemnite faunas have the same basic distribution as the ammonoids, with different families populating Boreal and Tethyan realms.

These distinct populations persist throughout the Cretaceous, although the generic character of the populations vary quite considerably. Forms such as *Belmnopsis* and *Hibolithes*, important in the Jurassic, declined in Cretaceous times, whilst *Mesohibolites* and *Neohibolites* expanded in both numbers and distribution. Throughout the Cretaceous the evolution of the belemnites continued, and was accompanied by numerous migrations. At the end of the Maastrichtian, however, the majority of belemnites became extinct, only the family Bayanoteuthidae, persisting in a limited fashion until the Upper Eocene. It is possible that the appearance of the voracious marine reptiles, the mosasaurs, was important in the extinction of belemnite stocks, though climatic changes could have been just as important. The problem of explaining the extinction of the belemnites is not made easier by the appearance of the true squids and octopuses in Mesozoic times. One is forced to ask; why should one group die out whilst two others of similar form survive and prosper?

Mesozoic arthropods, lacking the presence of the trilobites, are dominated by the crustaceans and insects. The crustaceans are again represented in many deposits by the ostracods, forms with ornamented shells being of importance in the Jurassic and Cretaceous. *Cypridea* and *Cythereis* occur throughout these periods, the first inhabiting freshwater environments, the second, marine. In certain sediments ostracods cover the bedding planes and within the famous Jurassic outcrop of Lulworth Cove, Dorset, are associated with salt pseudomorphs, an association indicative of somewhat fetid conditions of deposition.

Eryon arctiformus, a Jurassic crustacean from the Upper Jurassic, Solnhofen, Germany 8·4 cm (3·3 ins) long.

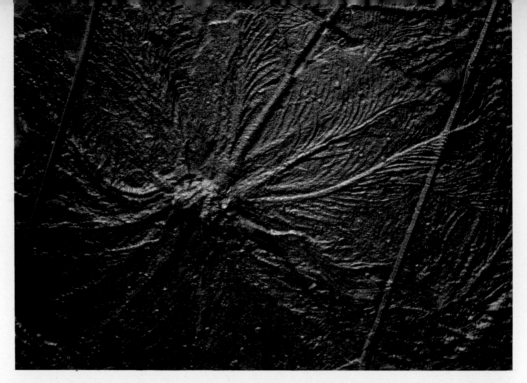

The arms and stems of a crinoid from the Lower Lias of Wittenburg, Germany.

True crabs and lobsters appear in the Jurassic and are amongst the last malacostracan crustaceans to appear in the fossil record, the ancestors appearing first in Triassic times. Abundant malacostracans are found in the Upper Jurassic sediments of northern Germany. They are associated with numerous chelicerates, with the so-called 'sword-tails' or king crabs being most common. Fossils of branchiopods and barnacles are discovered in the Mesozoic, but although their history goes back to early Palaeozoic times they are of limited importance.

Amongst the insects, winged forms became important on land, with the cockroaches, beetles, lacewings, moths, bees, wasps, ants and bugs either appearing or expanding during Mesozoic times. The information provided from the detailed collection of fossil insects indicates that their habits have changed little since their appearance. Insect faunas are, however, useful in the determination of ancient climates and microenvironments.

Various echinoderm groups of the Palaeozoic are still represented in the Mesozoic. Of the so-called 'fixed forms' only the crinoids survive, the vast majority of echinoderms being asterozoans, echinoids and holothurians.

Crinoid representation in the Mesozoic is maintained through the presence of the subclass Articulata, all other groups having died out during the late Palaeozoic. Articulates have a dorsal cup which is somewhat reduced in size, with fewer plates present above the radials. The first articulates appear in the Lower Triassic, their ancestry being traced back to the Palaeozoic inadunates. Stemmed articulates flourish in the Triassic, with some thirty genera being recorded from Europe and the Americas. *Pentacrinites* and *Isocrinus* are long-lived genera which appear in the Triassic. They are characterized by large crowns and a large number of branching arms.

Of the 250 crinoid genera known from the Jurassic, 10 per cent are stemless, having abandoned the stem at an early larval stage. *Saccocoma* from the Jurassic, although small, had wing-like expansions on the arm plates, which probably helped in swimming and food collection.

In Cretaceous times articulate stocks declined to less than 200 genera but individual species were locally abundant. *Marsupites*, a large-plated stemless form is used in the zonation of the Upper Cretaceous, whilst *Pentacrinites* and *Isocrinus* have continued representation in Cretaceous deposits.

Unlike the crinoids, echinoid numbers and diversity increase dramatically in Mesozoic times. From cidaroid ancestors, seventeen of the eighteen orders of the subclass Euechinoidea evolved during the Mesozoic. As with other phyla, the Triassic record of the echinoids is poor, with only six families having definite representation. No early Triassic forms are known and those of later times are generally small, geographically isolated and rather cidaroid in appearance. In fact, one order is called the Hemicidaroida.

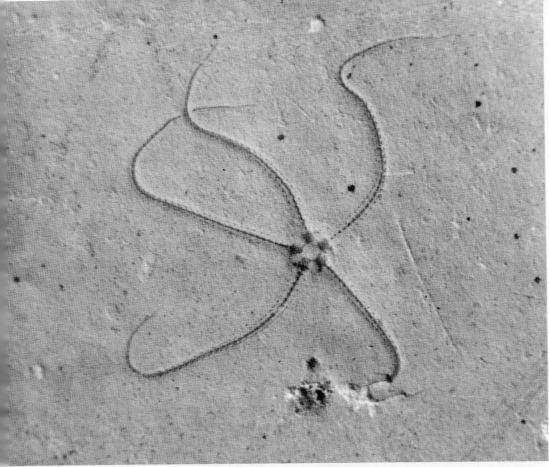

Some Jurassic echinoids such as *Hemicidaris* bear a strong resemblance to their Palaeozoic ancestors. 4·1 cm (1·6 ins) in diameter.

A brittle star from the Lithographic limestone, Jurassic, Solnhofen, Germany.

The echinoid *Micraster*, various species of which are used in the subdivision of the Upper Chalk, Upper Cretaceous. 5 cm (2 ins) across.

In the Lower Jurassic, numerous families appear in a burst of echinoid evolution. The structure of the skeleton stabilized in early Mesozoic times, with two rows of plates in each of the plate areas. Most early forms had the mouth and anus in the central position, but throughout the Jurassic and Cretaceous numerous forms arose in which this regular condition had been modified. In these the so-called 'irregular echinoids' the anus had migrated out of the circlet of apical plates. This migration, backwards and ventrally, results in loss of radial symmetry. Bilateral symmetry evolved in many family lines.

The change in the form of the test often results in a well-defined orientation which indicates the development of locomotion in a preferred direction. In some irregular echinoids the pored plate areas become shortened and petal-like and many species adopt a burrowing mode of life. Flattened echinoids develop as in the Palaeozoic 'regulars', with food grooves helping in the feeding process. Echinoids are gregarious animals and it is not unusual to find many hundreds in small, localized areas.

Both regular and irregular echinoids occur throughout the Jurassic and Cretaceous. *Palaeopedina* and *Phymosoma* are characteristic of Jurassic regular echinoids, whilst *Pygaster* and *Clypeus* are irregular types of the same period. In the Cretaceous, numerous families of irregulars dominate echinoid communities. Amongst these are the rather flat clypeastroids, and the heart urchins (spatangoids), both of which are important stratigraphically, the last being of particular significance in the Upper Cretaceous. *Micraster* (Spatangoida), *Holaster* and *Echinocorys* (Holasteroida) are very important Cretaceous genera.

The numbers of asterozoans preserved in fossil record of the Mesozoic are limited. Both star and feathery varieties are known but discoveries are extremely rare.

Amongst the vertebrates, the bony fishes and reptiles were the dominant Mesozoic groups. The ray-finned fish outnumber both lobe-finned and cartilaginous varieties. Early Actinopterygii (ray-finned bony fish) had heavy scales and mainly cartilaginous skeletons. From these the holostean actinoptergians evolve in Triassic times, having thinner scales and a reduced fin skeleton. *Caturus*, a Mesozoic holostean, and related forms appear intermediate between the heavy scaled chondrosteans and the most advanced actinopterygian group, the teleosts. In these the lung is completely transformed into an air-bladder and jaw mechanism such that when the jaws are protruded towards the prey, the mouth cavity enlarges to suck in large quantities of water. The body of a teleost is mainly composed of powerful swimming muscles. Teleosts and advanced teleosts evolved in the Jurassic and Cretaceous respectively.

Clypeus ploti, a rather flat, bottom-dwelling echinoid from the Jurassic, Gloucestershire, England. 5·7 cm (2·2 ins) in diameter.

A rare find, *Caturus*, a Jurassic holostean fish, choked by its prey. Maximum length 156 cm (60·8 ins).

Cyclobatis, a ray from the Upper Cretaceous of Syria. Width of block 18 cm (7·1 ins).

A bedding plane with the remains of an icthyosaur and a fish; Jurassic, Germany. Ichthyosaur length 1 m (39 ins).

Far right
Lariosaurus, a Triassic aquatic reptile, related to plesiosaurs. 23 cm (9 ins) long.

Sharks and rays represent the cartilaginous fish throughout the Mesozoic. In general, these are conservative stocks, with long-ranging genera like *Hybodus*, *Myliobatis* and *Lamna* occurring quite frequently.

Bony fish fossils are usually rare, but occasionally thousands of individuals are discovered in a mass grave. One such discovery is the Lebanese fish bed.

The fish were not the only marine vertebrates during the Mesozoic, for at various stages reptiles such as icthyosaurs, plesiosaurs, turtles and mosasaurs returned to the sea.

On land the story was more complex, with appearances and extinctions presenting a complex evolutionary mosaic. By Triassic times reptiles of the anapsid (skull without temporal openings) type were of limited importance, for many of the Palaeozoic stem reptiles had died out due possibly to competition or attack from more advanced groups. From anapsid stocks the turtles (anapsid), icthyosaurs, plesiosaurs and placodonts evolved. The last three groups were euryapsid reptiles.

Whilst the anapsids failed in the Late Palaeozoic, the mammal-like reptiles (synapsids) increased in both numbers and diversity. At the end of the Permian the sail-back lizards such as *Dimetrodon* and *Edaphosaurus* (pelycosaurs), gave way to the highly successful therapsids. In the Lower and Middle Triassic this group dominated terrestrial reptile communities. Skeletons of various individuals indicate that several groups evolved, each showing advances over the pelycosaur stocks. Therapsid skulls show improvements in dentition, an increase in brain size and in some late forms the development of a secondary palate. Both carnivorous and herbivorous therapsids are known.

In Late Triassic times the reptile world changed dramatically, the therapsids giving way to the thecodont archosaurs. The latter are early representatives of the most advanced group of reptiles, all of which have a diapsid skull (skull with two temporal openings). The thecodonts, or 'tooth in socket' reptiles, had an improved limb structure and, as their name suggests, an advanced dentition. They were carnivores, and many were larger than their therapsid rivals. It is probable that these animals annihilated therapsid stocks and replaced them as the dominant reptile stock.

This dominance was short-lived as the Upper Triassic also saw the establishment of the dinosaurs, crocodiles and pterosaurs, three groups which were to dominate vertebrate faunas throughout the Jurassic and Cretaceous periods. Mammals evolved from the advanced therapsids in late Triassic times, but their influence on vertebrate life was relatively unimportant until the dawn of the Cainozoic. Throughout the Mesozoic the mammals were small, rather delicate creatures, living as scavengers or insectivores, probably nocturnally. Carnivorous and herbivorous dinosaurs, often of gigantic size, restricted the development of mammalian stocks, whilst the pterosaurs overshadowed the appearance and evolution of the Mesozoic birds.

The fossil record of the dinosaurs and pterosaurs is well documented, with numerous beautiful discoveries from Jurassic and Cretaceous deposits. Often palaeontologists unearth complete fossils which provide invaluable clues to the body structure and mode of life of these animals. At the present time the problems how pterosaurs flew and whether dinosaurs were warm-blooded or not dominate the world of vertebrate palaeontologists.

As with other groups, the dinosaurs and pterosaurs die out at the end of the Mesozoic, their demise being linked, as before, with many possible changes or catastrophes. Climatic change alone is not sufficient to explain the extinction of either of these groups or that of many marine reptiles, ammonites, belemnites, rudists and planktonic foraminifera. However, climate linked with changes in vegetation and the replacement of one animal by another more suited to the new environment may provide the answers. In the Mesozoic, and the Cretaceous in particular, the flowering plants expand rapidly, with oak, beech, magnolia, willow, plane and poplar recorded from Cretaceous sediments.

At the end of the Cretaceous, the break-up of continents begun in late Triassic times, was almost complete, although the Atlantic and Indian Oceans had not attained their present widths and the continental masses were still moving. Continental drift led to animals and plants becoming isolated and this may have contributed to the extinction of several groups. The fragmentation of the great land masses of the Palaeozoic and early Mesozoic also provided large numbers of new ecological niches. Mountain building and the movement of continents into the polar regions can be linked with a changing climate throughout the early Cainozoic. In the absence of the dinosaurs, the warm-blooded mammals were to spread rapidly. With the disappearance of the ammonites, marine invertebrate communities were to be dominated by bivalves and gastropods.

Mud cracks and the footprints of the Triassic reptile *Cheirotherium* indicate that drought conditions were common 210 million years ago. Larger footprint 17·5 cm (6·8 ins) long.

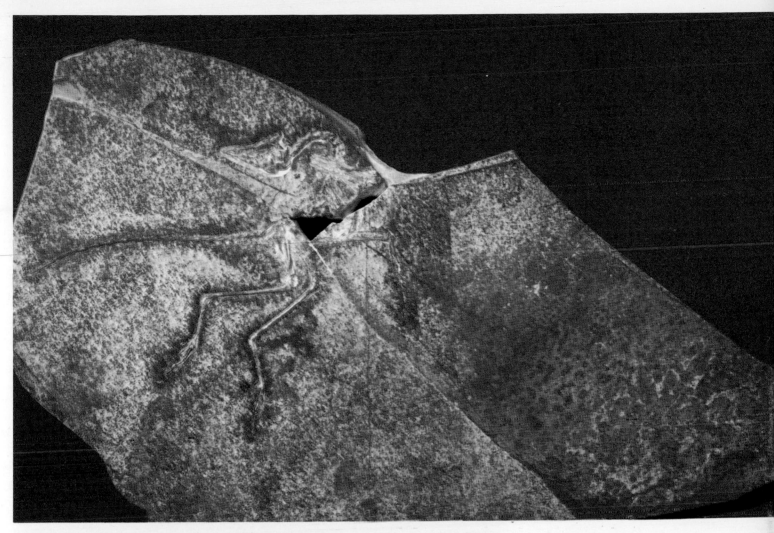

The most recent skeleton of *Archaeopteryx lithographica*, the earliest bird. Lithographic limestone, Jurassic, Solnhofen, Germany. Skull length 4·2 cm (1·6 ins).

The cone of the pine *Taxodiaceous* from the Cretaceous of Patagonia, South America. Width of cone 4 cm (1·6 ins).

The Cainozoic era

The Cainozoic era is divided into the Tertiary and Quaternary periods, which cover the last sixty-five million years of geological time. The Andes and Rockies had appeared by the end of the Cretaceous and the final episodes of the Alpine and Himalayan movements took place in the Upper Tertiary.

In the seas and basins of the Tertiary the marine faunas were very similar to those of the present day, although the Foraminiferida were represented by the large calcareous nummulitids, alveolinids and orbitoids. Some genera are widespread, others restricted to protected shelf environments. The coiled nummulitids have many chambers and some species reach several centimetres in diameter. During the Eocene several species of the genus *Nummulites* are used in stratigraphic correlation. The rapid spread of successive faunas appears to be generated from the Tethys region. Alveolinids and orbitoids are also good indicators of age and, like the nummulitids, often contribute to rock formation. By the Oligocene most of the earlier Cainozoic (Palaeocene and Eocene) forms had become extinct and shelf habitats were ready for a new fauna. By Middle Oligocene modern faunas had begun to appear, although the last few nummulites were quite common.

Tertiary planktonic faunas were characterized by a number of genera including, and related to, *Globigerina*. Algal skeletal elements also occur in fine-grained Tertiary sediments. The calcareous units of coccoliths, discoasters and rhabdoliths are extremely useful in petroleum geology. As planktonic organisms they provide a detailed time sequence for Tertiary distributions.

Tertiary sponge faunas are very similar to those of the Quaternary and Recent. Many of the genera important in the Cretaceous had died out by Tertiary times. Amongst those that persist is the boring sponge *Entobia*, known in abundance in the European Cretaceous through the species *Entobia cretacea*. *Entobia* is a very long-ranging form, being widespread in distribution, and alive at the present day.

Of the scleractine corals only a limited number of families became extinct in the Upper Cretaceous and early Tertiary, and coral communities have changed little, in family representation, during the last sixty-five million years. Certain genera and species were restricted in time and distribution, however, and forms such as *Turbinolia dixoni* and *Goniopora websteri*, both from the Bracklesham Beds of Sussex, England, are examples of corals restricted to a particular sediment.

The stony corals are always pre-eminent in any discussion on coelenterates. One should remember, however, that, as with the stromatoporoids in the Palaeozoic, the millepores (Hydrozoa) are of considerable importance in the construction of Cainozoic and Recent reefs.

Cheilostome bryozoans abound in the Cainozoic seas of Europe, North America and southern Australia. As a group they continue to expand, with numerous species recorded from littoral, sub-littoral and deep-water environments. Encrusting and delicate branching forms of exquisite beauty are

Recent diatoms. ×200

A polished section of an Oligocene algal colony, often termed a 'chou-fleur' colony. 12 cm (4·7 ins) wide.

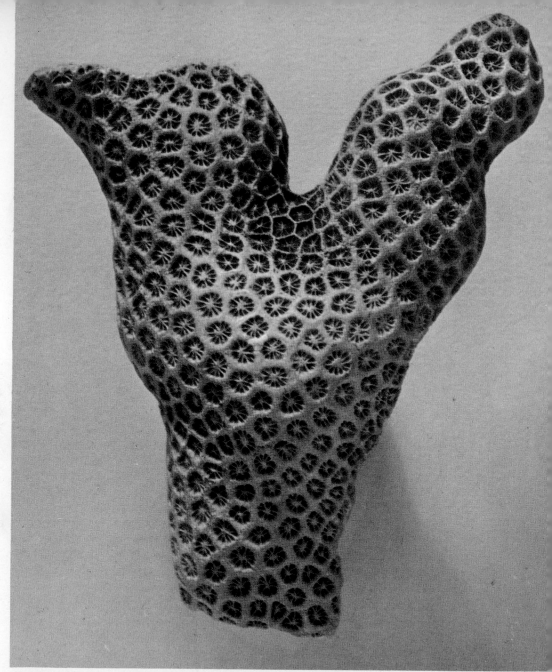

A scleractine coral *Septastraea
forbesi*, Miocene, Maryland,
Ohio, USA. 28 cm (11 ins)
long.

Fabellum candeanum, a
solitary hexacoral, Miocene,
Victoria, Australia. 2·5 cm
(1 in) wide.

Goniopora websteri, a hexacoral known only from the Bracklesham Beds, Middle Eocene, Sussex, England. Corallite diameter 5·8 mm (0·2 ins).

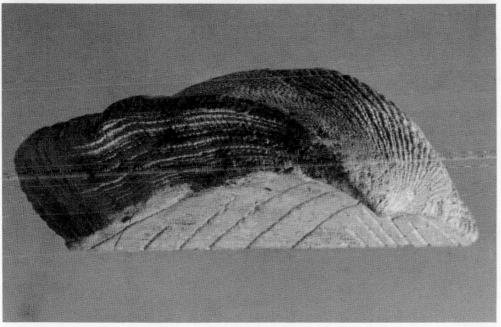

Arca biangula, an ark shell from the Eocene. *Arca* has a worldwide distribution; range Jurassic to Recent. 3 cm (1·2 ins) long.

commonly found amongst the strandline debris on modern beaches. Numerous cheilostome faunas have been described from the Cainozoic and, of these, that of the Coralline Crag (Pliocene) of East Anglia, England, is worthy of particular note. Numerous genera and species representing six or more families have been recorded from this horizon, which fully reflect the beauty of colonial organisms from a shallow-water environment. Cyclostome bryozoans are of secondary importance throughout Cainozoic and Recent times.

Brachiopods play a limited role in Cainozoic faunas, their presence being dominated by the terebratellids and terebratulids. Rhynchonellids persist and, although locally abundant, are of limited stratigraphic importance, as are the inarticulates and the articulate Strophomenida which are represented by two living genera.

Molluscs assumed a modern aspect in the Cainozoic era, with large numbers of new genera and species evolving in several distinct provinces. Pectens, oysters, cockles and clams flourish in the absence of the rudists. The trigonids, so abundant in early times, become restricted to the seas of the Australian archipelago, reducing in number until today they are represented there by a single genus.

Of the pectens, some, like *Amussiopecten* of the Miocene, are of value as index fossils. Amongst the oysters, forms similar to *Gryphaea* reappear, but geographically they are restricted to North Africa and central Asia. Clams such as *Arctica* are thought to indicate the migration of northern Boreal fauna southwards in Tertiary times, whilst the advanced clam *Venericardia* remains, as in the Cretaceous, a cosmopolitan genus. Other heterodonts related to the living cockle *Cerastoderma* occur in large numbers in early Cainozoic faunas, but *Cerastoderma*, like the living scallop *Pecten*, does not appear until the Oligocene.

The heterodontid, giant clam *Tridacna*, which is often found in association with coral reefs, evolved in the Tertiary. Of the various species referred to the genus, *Tridacna gigas* is known to weigh over 112 kilograms (250 pounds) and reach 1 metre (3·3 feet) in length. Boring bivalves such as *Teredo* and the species *Tridacna crocea* abound in the Cainozoic.

The reorganization and modernization of molluscs begun in the late Mesozoic is clearly shown in Cainozoic communities. Palaeozoic to early Cretaceous gastropods consisted mainly of herbivores or sediment feeders. Late Mesozoic and Cainozoic groups include predators, scavengers and

A fossil scallop from the Oligocene of Libya. 4 cm (1·5 ins) high.

A fossil oyster from the Miocene of Libya. 18 cm (7 ins) high.

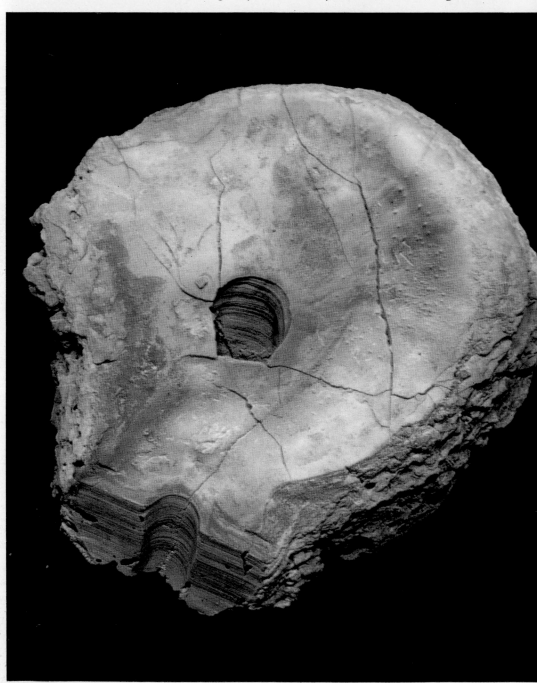

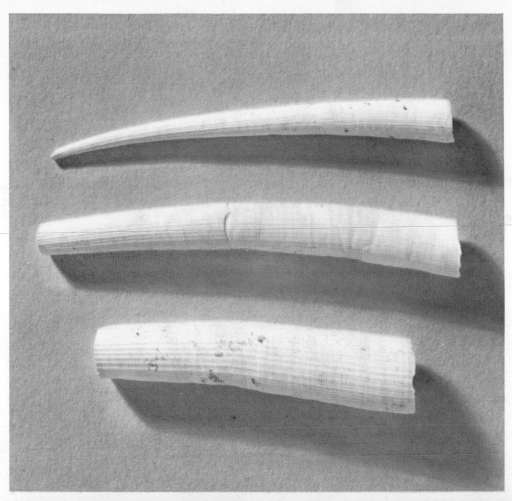

Antalis striata, a tusk shell
from the Eocene, France.
4·2 cm (1·6 ins) long.

Mollusc, a recent chiton.
9·5 cm (3·7 ins) long.

Natica, a predatory gastropod, ranges from the Triassic to Recent, with worldwide distribution. 2·1 cm (0·8 ins) long.

The bivalve, *Teredina personata*, is known from borings in wood, like these from the London Clay, Basingstoke, England. Average diameter of borings 8 mm (0·3 ins).

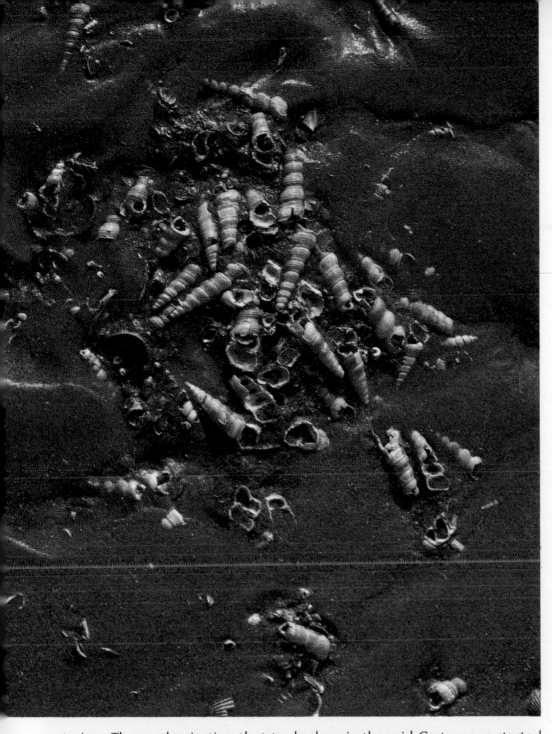

Shell drift of *Turritella sulcifera* Eocene, Bracklesham Bay, England. Individual specimens 5 cm (2 ins) long.

parasites. The modernization that took place in the mid-Cretaceous started in the northerly temperate latitudes, and spread south into the more conservative tropical communities. It is possible that the extinction of the ammonites and certain vagrant echinoids encouraged the development of these diverse feeding types. Holes drilled by *Natica* and other predatory gastropods are a common and interesting feature in many Cainozoic assemblages. Gastropods reach their greatest diversity and abundance in Cainozoic deposits, with freshwater and terrestrial forms becoming common. *Viviparus*, *Planorbis* and *Lymnaea*, all freshwater gastropods, are of value in late Cainozoic stratigraphy.

Gastropods are also very useful as indicators of ancient climates and geography. The species *Velates schmeideli*, extant today in the Indian Ocean, is found as a fossil in the Lower Tertiary of the Paris basin. Numerous species of *Turritella* and *Littorina* are characteristic of Cainozoic shelf faunas.

Post-Cretaceous nautiloids are represented by two lineages; the first, represented by *Aturia*, having a highly fluted septa. The other, of which the modern *Nautilus* is a member, is characterized by rather simple, sinuous septa. Until late Miocene times the nautiloids were quite common, but with the extinction of *Aturia* and its relatives their numbers and importance were reduced to their present position of relative insignificance. Since the Cretaceous, soft-bodied cephalopods have expanded quite considerably, with octopuses, squids and cuttlefish forming great swarms in some seas and oceans. Of these the octopuses may be the descendants of the Cretaceous ammonites, with the vast majority of forms having lost the characteristic shell. Only the female of the genus *Argonauta* produces a shell to receive and transport its eggs. Unlike the shells of ammonoids, however, this brood pouch lacks septa.

The record of cephalopods in Tertiary and Quaternary deposits is largely comprised of squid remains. The soft bodies of octopuses are rarely preserved in the fossil record.

It is probable that the gradual cooling that has taken place since the end of the early Tertiary has affected the distribution and abundance of certain plant and animal stocks. However, temperature is only one of a number of factors that influence stocks. Like many other groups, the Cainozoic ostracods are also affected by depth, energy and salinity. Depth is usually associated with a lack of diversity, higher energy with stronger carapaces and decreased salinity with specialized forms. Certain forms of ostracods are associated with certain types of sediment, and make excellent index fossils. Their role in the study of Cainozoic stratigraphy increases daily. A detailed study by palaeontologists of the genus *Saida* has revealed that after evolving in Europe in the Cretaceous, the genus migrated to Australasia in the Lower Cainozoic. *Saida* persists in the Southern Hemisphere to this day, the last European discoveries being from rocks of Lower Cainozoic age.

In the Tertiary deposits, the record of the insects and crustaceans increases considerably. Spiders, poorly known in the Mesozoic, are really abundant in Tertiary deposits and, together with winged insects, are especially well-preserved in early Tertiary amber from the Baltic region of Europe. Beetles also reflect the increased importance of insects in the Cainozoic, the numbers of species increasing from 300 in the Cretaceous to 2 300 in the Tertiary and 300 000 at the present day. This dramatic diversification is probably linked with the continued spread of flowering plants and the rise of the mammals.

A well-preserved beetle from the Lower Eocene lake deposits of Menat, Central France. 1·4 cm (0·5 ins) long.

Ilia nucleus, from post Tertiary deposits, Philippine Islands. This is the 'Medicine Crab' of Chinese druggists. Individual 2 cm (0·8 ins) wide.

Palaeocarpilus aquilinus, a crab from the Eocene of Libya. 14·2 cm (5·5 ins) wide.

Crab, *Portunus lancetidactylus*, from the Oligocene of northern Caucasus, USSR. 7·8 cm (3 ins) wide.

Of the crustaceans, crabs assume a special stratigraphic role in the Tertiaries of the Gulf Coast and Pacific border of North America. *Callianassa* is one of the crabs used as a zone fossil in these areas, and its burrows are found as trace fossils. Numerous crabs and lobsters have been collected from the London Clay (Eocene) of the Isle of Sheppey, England. *Balanus*, the living barnacle, is found as a fossil throughout the Cainozoic.

The echinoids of post-Cretaceous times show considerable changes from earlier forms. All major orders are represented throughout the Cainozoic but the relative importance of the various families and superfamilies differs significantly.

The Echinacea show considerable diversity, with individual families showing a preference for different environments. Temnopleurids like *Paradoxechinus* from the Eocene of Europe and the Oligocene of Australasia prefer deep water, but members of the families Echinidae and Strongylocentrotidae abound in waters of the littoral zone. The low percentage of forms occurring in turbulent regions preserved in the fossil record is misleading, and it is more than likely that they were as common as deep-water species in Tertiary deposits.

The sand dollars, Clypeasteroida, reach a maximum of thirty-six genera in the Miocene. They exist either on or beneath the surface of the substrate. The rapid evolution of ambulacra, coupled with development of food grooves, makes them stratigraphically important in the Cainozoic.

An irregular echinoid
Eupatagus from the Oligocene
of Libya. 5·7 cm (2·2 ins)
long.

A Pliocene barnacle, *Balanus*
(*Chiroma*) from Mombasa.
Individual 2·9 cm (1·1 ins)
long.

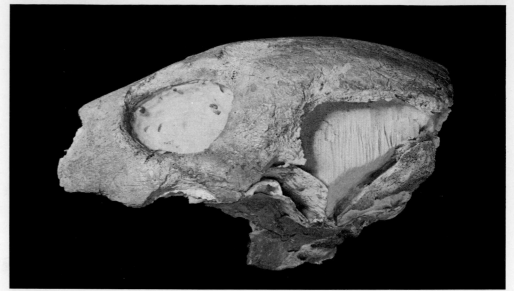

The skull of a fossil sea turtle from the Eocene phosphates of Tunisia. 19·2 cm (7·5 ins) long.

Teeth of the Eocene shark *Odontaspis*. 2 cm (0·8 ins) long.

Priscacaria, a fossil teleost fish from the Green River formation, Wyoming, North America. 13 cm (5 ins) long.

The spatangoids and cassiduloids also show considerable expansions during the Tertiary. Spatangoids flourish during the Eocene, with the appearance of a maximum number of species. Since then the group has undergone a steady decline. The post-Tertiary decline of the cassiduloids, however, was anything but steady, with a reduction in generic numbers from 500 during the Tertiary to sixteen at the present time.

It would appear that many of the successful Cretaceous stocks failed during the Cainozoic era, the presence of only one living genus of the important family Holasteridae tending to support this hypothesis.

As in the Cretaceous, Cainozoic crinoids all belong to the subclass Articulata. At first the group is represented by only a few genera, but in the late Pliocene a considerable expansion takes place. The free-swimming forms, the comatulids, are the dominant crinoids in modern seas, with some ninety genera recorded.

Starfish discoveries are not very common in Cainozoic deposits, although exceptional discoveries have been recorded from the London Clay and the Bracklesham Beds of England.

If the Mesozoic was the 'age of the reptiles', then the Cainozoic is justly identified as the 'age of mammals', for, in the absence of the dinosaurs and the great marine reptiles, the shy, nocturnal animals of the Mesozoic were soon to become rulers of the world. Their rise was linked with the final break-up of the continents and paralleled by the diversification of the flowering plants.

The age of mammals began some sixty-five million years ago with the dawn of the Palaeocene. Free to occupy the niches vacated by the reptiles, the mammals underwent one of the greatest radiations known in the fossil record. In the warm humid climate of the Palaeocene, mammals such as *Ptilodus* and *Taeniolabis* marked the end of the ancient order of the multituberculates, whilst marsupials and placental mammals heralded the rise of new, more advanced, orders. Carnivores and herbivores were common, and primitive hoofed animals appeared for the first time. Primates were also evident in late Cretaceous and early Palaeocene faunas of Purgatory Hill, eastern Montana.

Right
Merycoidodon, a plant-eating mammal related to cattle and pigs. Oligocene, Dakota, USA. Skull 14 cm (5·5 ins) long.

Below
Phororhacos longissimus, a Miocene, flightless bird from South America. Skull 46 cm (18 ins) long.

Far right
Diaphorapteryx hawkinsi, a flightless rail from the late Pleistocene of New Zealand. 52 cm (20 ins) high.

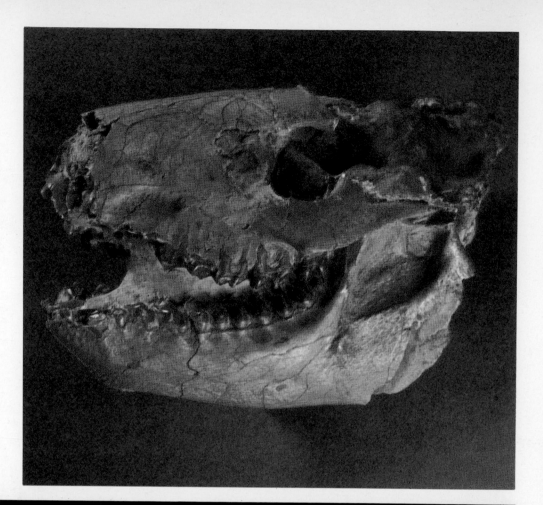

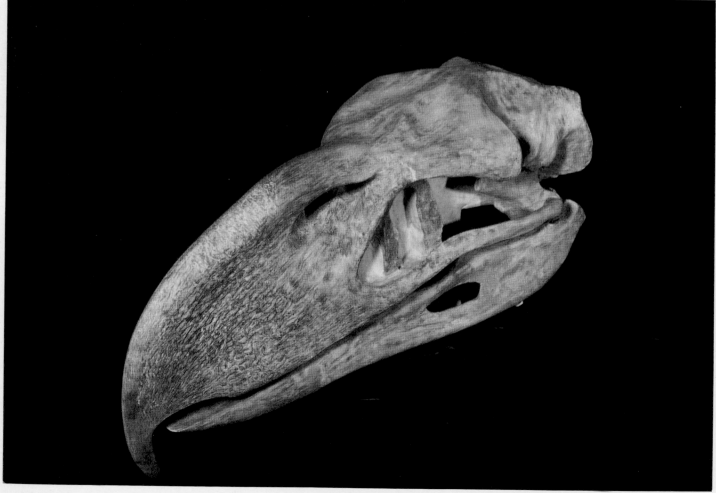

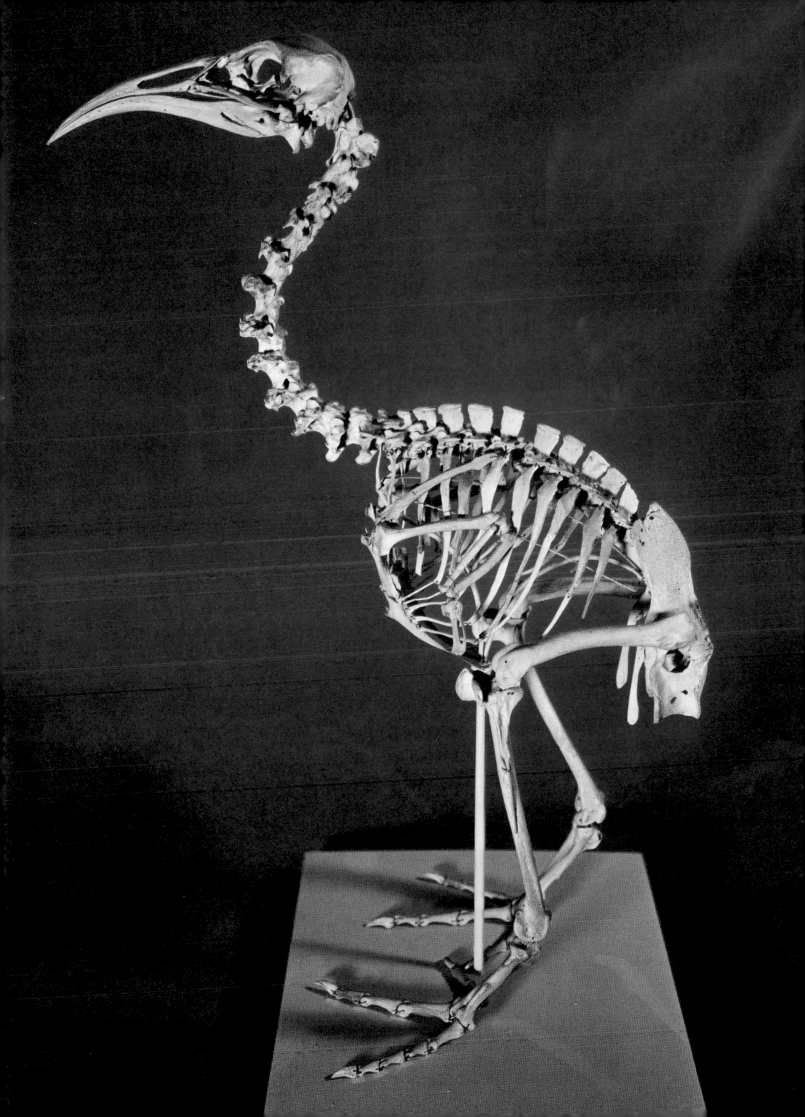

In the Upper Palaeocene, the first true rodent appears in the trees, a squirrel-like form called *Paramys*. On the ground ancestors of the sloths and anteaters appear, together with herds of hoofed animals and true carnivores.

Crocodiles, alligators and soft-shelled turtles flourish in the swamps and rivers, and giant sea turtles roam the seas. Birds occupy the niches formerly held by the pterosaurs and one group, the giant terror cranes, including *Gastornis*, attempt to maintain the dominance once held by the two-footed dinosaurs.

From sediments of Eocene times, mammalian fossils provide further information of the great diversification. From humble beginnings the early horses and even-toed ungulates showed a progressive increase in size. The evolution of the horse is one of the best documented stories of the fossil record, with the changes in overall size, dentition and limbs indicating a movement from forest browsers to fast-running grazers.

The first bats also appear in the Eocene and one concludes that, in general, mammals, like the birds, achieve a modern aspect in this period. In the seas the first whales, the Archaeoceti, show a rapid adaptation, and by late Eocene the zeuglodont whales such as *Basilosaurus* have reached over 15 metres (50 feet) in length. Their relative abundance in the Eocene and Miocene is often attributed to an abundance of soft-bodied cephalopods.

A tooth of the elephant-like *Dinotherium* from the Lower Pliocene of Germany. Width across crowns 6 cm (2·3 ins).

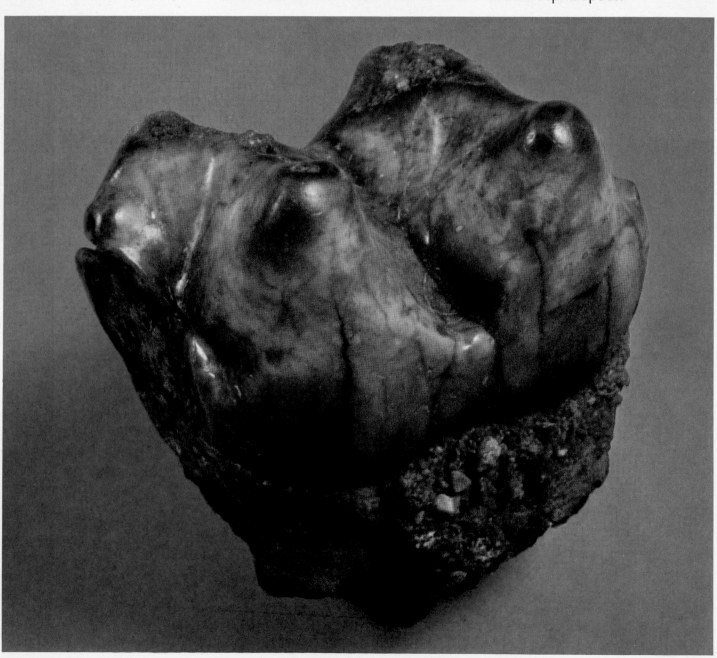

The leaf of a flowering plant from Ypresian lake deposits of Menat, Central France. 3·2 cm (1·2 ins) long.

Elephants, rhinos, dogs and cats all appear in the late Eocene. Their numbers increase throughout the Oligocene and Miocene and many large and spectacular species have been recorded.

From the Eocene onwards the story of the mammals is one of experimentation, with numerous short-lived offshoots representing attempts at solving particular problems. The number and variety of groups appears endless, reaching a significant peak in the Miocene with the appearance of man's ancestral cousins. Human ancestry rests amongst the Old World primates and the clues to its development and success are still being unravelled from the sediments of Africa and Asia.

As stated previously, the evolution of the plants, in part, paralleled the diversification of the mammals. Luxuriant fossil forests in Iceland, Greenland and Spitzbergen prove that the Palaeocene was warm and humid. Willows, poplars, elms, ivy and palms are all recorded from the Palaeocene, whilst the seeds of *Nipa* and *Oncoba* from the Eocene of the Isle of Sheppey draw comparison with the modern tropical forests of the Malay Archipelago.

The Florissant flora (Oligocene) of Colorado suggests drier, more temperate conditions, with oak, spruce and poplar amongst the numerous genera recorded. European Oligocene floras also indicate cooling in northern areas, although residual pockets of more tropical plants persist locally.

The final break-up of the continents in the Cainozoic, coupled with the building of the Alps, Atlas Mountains and Himalayas, obviously affected the climate. The final results of these great geological phenomena are seen today in the distribution of plants and animals and in the preservation of the archaic marsupials in the protected niches of Australia and South America.

Trace fossils

Throughout the fossil record, sedimentary structures occur which are the result of biological activity. These structures are accepted as true fossils and their study, palichnology, is recognized as a very important part of palaeontology. Unlike 'body' fossils, traces are often improved by the effects of chemicals and compaction.

It is often difficult to associate traces with the animals or plants that created them and, thus, application of a systematic approach to identification is fraught with difficulties. One animal may manufacture a variety of traces during its 'working' life and different animals having the same feeding habits may leave the same trace. However, the geometry of traces is controlled by the depositional environment, and subdivision of structures within environments offers an acceptable method of classification.

Many traces are the result of the activity of soft-bodied creatures, such as the annelid worms. These segmented animals are poorly represented in the fossil record for only a few have hard parts in the form of jaws or opercula. Apart from rare finds of impressions, like those of several genera in the Burgess shale, the record of worms is restricted to trails, tracks, burrows and castings, examples of which date back to the Precambrian era.

Worm traces, like those of other organisms, often reflect function, and since the early 1950s palaeontologists have recognized five groups of functional traces. These are:

Repichnia. Trails or burrows formed by active benthonic animals moving in a determined direction.

Pascichnia. Winding trails or burrows, of active sediment eaters, formed when the creatures concerned 'graze' a certain surface area avoiding double coverage.

Fodinichnia. Burrows created by hemisessile deposit feeders, which act as a permanent retreat shelter during feeding forays.

Domichnia. The permanent shelters of active or hemisessile animals.

Cubichnia. Shallow resting tracks corresponding to the outline of the animal; formed when active animals hide or rest in sediment (after Seilacher 1953).

These functional traces are responses to environmental conditions, and as certain types of sediments generally indicate a particular environment, it follows that specific trace fossils will be restricted to certain sedimentary environments. Trace fossil communities can therefore be recognized and assigned to one of several major ichnofacies.

Cruziana tracks from the Ordovician near Salamanca, Spain. Area illustrated 1·5 m (58 ins) wide.

Thalassinoides burrows from the Cretaceous phosphates, Morocco. Single burrow length 8–12 cm (3·1–4·7 ins).

In general, the complexity of traces increases with depth, and the ichnofacies can be recognized by the relative abundance of one or more of the functional groupings listed previously. To animals that live in shallow-water environments, protection is very important and tracks and burrows of the cubichnid and domichnid type predominate. Shallow-water environments are here regarded as those of littoral and sublittoral depths; together they are termed the Cruziana facies. In the deeper zones, towards the edge of the continental shelf, deposit feeders become common, and in the fossil record their abundance is reflected by the increase of fodinichnia.

Footprint of a theropod dinosaur from Triassic sediments, Midelt, Morocco. 35 cm (13·6 ins) long.

The Zoophycos facies, sublittoral to bathyal, is characterized by the dominance of complicated fodinichnia. In deeper water environments, organisms gain little by hiding within silts or muds and, due to the scarcity of food, spend most of their time feeding. Their search is systematic and results in small, looped and septate traces.

In the Nereites facies, bathyal environments, pascichnia of active deposit feeders dominate. The traces are complex, with meandering and spiral trails representing the activities of organisms which lived at great depths.

Trace fossils are very useful as indicators, providing a great deal of information on depth and environments. They also provide information on rates of sedimentation, and in some instances can be used as index fossils. It is also possible to reconstruct the habits of various fossil groups through the study of their traces.

Traces of vertebrate animals include the burrows of Devonian lungfish and the footprints of Mesozoic dinosaurs. Again, they provide the palaeontologist with additional information on the life-style of a given animal, such as enabling the speed at which the dinosaurs travelled to be calculated.

Fossils and man

From the information set out in previous chapters it is obvious that fossils are of considerable importance to both the academic and professional palaeontologist. They provide information on the appearance and structure of past communities, and reveal the evolutionary changes that have taken place in the plant and animal kingdoms throughout time. Numerous groups show rapid evolutionary development, and species that are both widespread and short-lived are used as zone fossils.

The organic nature of fossils was recognized by both Xenophanes and Leonardo da Vinci (1452–1519), but only in the last 200 years or so has this suggestion been accepted and fossils used in both the correlation of strata and as clues in the concept of organic evolution.

William Smith (1769–1839), known as the 'father of geology', was the first person to recognize that different strata were deposited, in sequence, one upon another. He also recognized that fossils were common to various horizons throughout the land, and through methodical collection and accurate recording was able to use fossils for correlation. Robert Hooke (1635–1703) had suggested this over a century earlier, but, unlike Smith, had failed to develop his ideas.

It is to their everlasting credit that the geological maps prepared by Smith, and his contemporaries Cuvier and Brogniart, bear comparison with the more detailed results of recent surveys. Since those early days geologists have studied the entire sequence of strata in many lands. In the geological mapping of a region and the compilation of its geological history, correlation is all important. The use of fossils to determine an orderly sequence of rocks and to recognize and correlate rocks which are separated geographically is termed biostratigraphy.

With the publication of *The Origin of Species* in 1859, Charles Darwin made an outstanding contribution to scientific thinking. His major theme, that of natural selection, was not entirely new but, like William Smith, Darwin had organized a great weight of convincing evidence to support his theory. Before Darwin, Georges Bufon (1707–1778) and Jean Baptiste de Lamarck (1774–1829) had written on an evolutionary process based on the inheritance of acquired characteristics. The academic climate of their day was wrong, however, and general support was withheld until Darwin had published his thesis.

In support of his thesis Darwin had accumulated evidence from numerous sources, including palaeontology. In the sixth edition of his book he discusses the imperfections of the fossil record and makes particular note of 'the absence of intermediate varieties at the present day'. This is obviously a reference to the absence of 'missing links' and common ancestors of present-day faunas and floras. Amongst the fossil evidence quoted by Darwin is the fossil sea-cow *Halitherium*, which, unlike living sea-cows, possessed traces of vestigial hind limbs and pelvis, a clue that suggested that the sea-cows had evolved from four-legged ancestors.

A spider in resin. 2 cm
(0·8 ins) wide.

Far right
A trilobite broach.

Fossil evidence remains a conclusive testimony in support of the concept of evolution. Before the study of genetics provided a key to the forces of heredity and variation, the fossil record was without doubt the major source of evidence in support of evolution. The documentation of horse evolution alone provides ample support for this statement.

The use of fossils by man to unravel the mysteries of stratigraphy and evolution are thus explained. Since those early days, great strides have taken place in palaeontological thinking. Fossils are no longer used simply for correlation; foraminiferids and ostracods have become the tools of the petroleum geologist, providing information not only on the age of core samples, but also on the sedimentary facies and the geologic setting of fuel-rich deposits. Spores and pollen are also used by geologists in the dating of sediments and the analysis of past environments and climates. Both spores and pollen have a decay resistant, outer wall which helps in their preservation. The record of spores dates back as far as the Silurian, whilst pollen analysis has been successful in the study of Upper Mesozoic rocks. Pollen grains quickly oxidize when exposed to the air, and so the criteria applied for the preservation of other organisms are equally important in the protection of these microscopic grains. Pollen grains are commonly associated with water-logged sediments, and peats, in particular, yield exceptional concentrations.

In order to estimate the populations of floras, sample pollen counts are made. When plotted on a graph these illustrate the ratios of different species at a given time, and allow the palaeontologist to note any change that has taken place. In more recent studies it has proved possible to plot the effects man has on a certain environment.

Problematic fossils such as conodonts, small jaw-like or toothed structures found in rocks of Ordovician to Triassic age, are of particular importance in Palaeozoic stratigraphy. Conodonts are composed of calcium phosphate, and although they have been associated with numerous phyla, their true zoological relationships are so far undetermined.

Apart from their value as stratigraphic and palaeoecological tools, fossils are also important in the economic sense. Fossil fuels such as coal, oil and natural gas are the result of the accumulation of and decay of organic remains.

The majority of coals are formed by the partial decay of plants that accumulated in swamps, protected from an oxidizing environment. The thickness of coal seams varies and they are frequently associated with root horizons and marine and non-marine fossil bands. Often these beds show distinct rhythms and together they are termed a coal series. The rank of a coal is related to the degree of metamorphism that has taken place since accumulation began. The classification of coals ranges from lignite to anthracite, with sub-bituminous and bituminous as the intermediate ranks. Peat is not regarded as a true coal, although it does form the base of a continuous series.

The tooth-like conodonts are among the more problematic fossils known to man. They have in turn been associated with gastropods as scraping organs or with other animals as grinding devices within the gut. 2 mm (0·08 in) long.

An oil rig in the North Sea. During the various stages of exploration, core samples are extracted for palaeontological investigation.

Seed ferns such as *Alethopteris* were abundant in the coal-forming swamps of the Coal Measures. Individual leaflets up to 4·2 cm (1·6 ins) in length.

The majority of bituminous and anthracite coals are found in Carboniferous basins. The great coal swamps of that period covered large areas of North America, Europe and Russia. In the Moscow basin, the coals of Carboniferous age were never buried beneath great layers of sediment and they remain lignitic in composition. Lignite deposits of economic importance are also found in the Eocene of West Germany.

Oil and natural gas are probably the result of the accumulation of silts with a high organic content. The remains of diatoms (plants) and zooplankton are rich in fats and proteins, and if trapped within sediments would undergo decomposition, in the absence of oxygen, to produce the various hydro-carbons. The generation of oil from organic matter is still poorly understood, but the investigations of palaeontologists and petroleum geologists include the study of both source and reservoir rocks. From these we have learnt that litte 'free' oil is found in rocks of freshwater origin and that the generation of oil and natural gas is a very slow process. Oil shales of freshwater origin occur in various parts of the world, including North America (Eocene) and France (Oligocene).

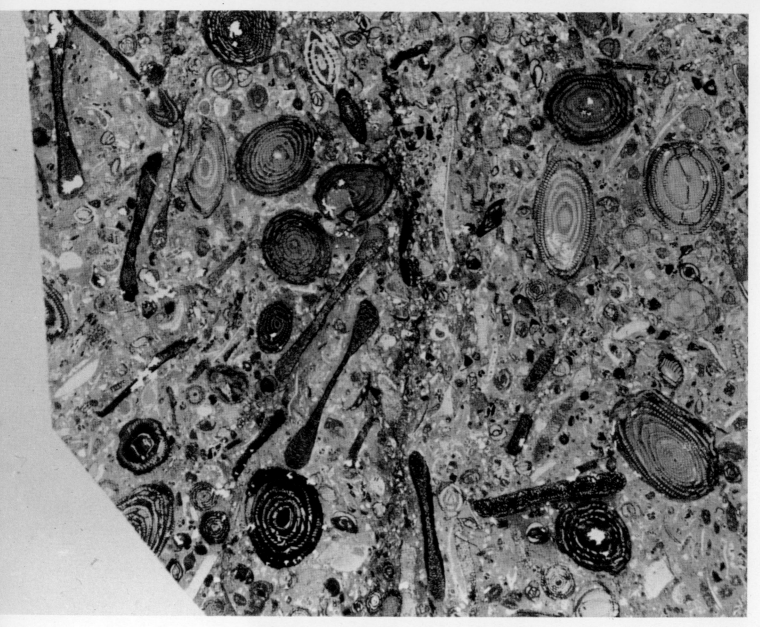

Alveolinid (foraminiferid) limestone from the Cainozoic of France. 3·6 cm (1·4 ins) wide.

Apart from the fossil fuels, many limestones and phosphates are the result of organic accumulations, and limestones and organic reefs may be associated with oil concentrations. Numerous groups of calcareous organisms have contributed to the formation of carbonate sediments throughout the geological record. Limestones may be composed almost entirely of microscopic foraminiferids or large bivalve shells. Economically, limestones are important in the construction and fertilizer industries. Phosphates, on the other hand, are associated with the accumulation of the skeletons of inarticulate brachiopods, trilobites and the vertebrates. In Morocco the vast phosphate deposits of Maastrichtian and Eocene age are packed with the remains of sharks, turtles and other vertebrates. Faecal pellets and coprolites of these animals also contribute to the high phosphatic concentration. The abundance of sharks in these beds provides some indication of the richness of the invertebrate foodstocks that existed during these times.

The seas in which the Moroccan phosphates were deposited bordered the opening Atlantic. Fossils and sediments indicate that this great ocean was non-existent in the early Mesozoic. The break-up of the great land mass called Pangea began over 200 million years ago. At first, a northern group of continents, Laurasia, split away from the southern conglomeration called Gondwana. Africa and South America then broke away from Antarctica and Australia, and after some sixty-five million years of drift we witness the opening of the south Atlantic. By late Cretaceous times the southern Atlantic was over

3 000 kilometres (1 900 miles) wide and the great ocean stretched from the Antarctic to the British Isles. The final rupture in the north took place in the Cainozoic, when the continents drifted into their present positions.

The story of continental drift is one which is supported by numerous lines of evidence, amongst which is a vast amount of palaeontological data. In many ways the theory of continental drift and plate tectonics has been the cornerstone of modern geology, with palaeontologists, sedimentologists, structural geologists and geophysicists working together towards a common goal.

The evidence provided by palaeontology supports the pattern of dislocation described above, for in late Palaeozoic times the *Glossopteris* flora is common to all the southern continents (Gondwana), whilst in the Tertiary the fragmentation of the two supercontinents appeared to stimulate the great radiation amongst mammalian stocks. In Mesozoic times the supercontinents appeared to have common reptilian communities but after the fragmentation of these great land masses in late Cretaceous and Cainozoic times, faunas characteristic of individual continents and island masses became apparent.

Deformed trilobites such as this *Angelina* can be used by structural geologists in order to analyse the amount of deformation that has taken place. 4·2 cm (1·6 ins) long.

Glossary

agglutinated foreign particles, such as sand grains, bound together by a cement secreted by the animal

anaerobic bacteria minute organisms that can exist in the absence of oxygen

bituminous shale fine-grained sediment, rich in oil-like minerals

body fossil preserved remains of skeleton or soft parts of animal

branching colony a colony consisting of a number of branch-like units

calcareous skeleton skeleton composed of calcium carbonate, $CaCO_3$

cast an impression obtained from a mould

chitin a complex organic substance made up of polysaccharides; forms external skeleton of graptolites and arthropods

chitinophosphatic combination of chitin and phosphate, found in external skeletons of certain brachiopods

crustal layer the outermost layer of the Earth, subdivided in two, sial and sima; sial 10–12 kilometres (6·2–7·4 miles) thick, sima 15–20 kilometres (9·3–12·4 miles) thick

Danian uppermost stage of the Cretaceous period

evolute successive chambers resting on the inner ones without any overlap, as in ammonites

facies those assemblages of fossils, structures or minerals that indicate the environment of deposition of a particular sediment

hemisessile term used to describe an animal which spends part of its life in one spot but which may move in order to migrate or feed

horizon geological time-plane recognized in rocks by means of fossils or a specific sediment

hydrospire respiratory structure in blastoids, developed parallel to ambulacral borders

ichnofacies facies characterized by the presence of a particular group of trace fossils, *see* facies

imperforate lacking pores

intertidal zone area between high and low tides

involute the outer chambers of a shell overlapping those inside, as in ammonites

isotope one of two or more forms of a given element, each having a different atomic weight

keratin organic, horny substance, found, for example, in some sponges

lamellate (colony) layers of skeletal material deposited one upon another

laminated layered structure

Lithographic limestone name applied to fine-grained carbonate rock used in print industry, the classic example being the Lithographic limestone of Bavaria, Germany

magma molten fluid formed in Earth's crust, which may solidify to give igneous rock

massive colony large group of closely packed individual skeletons, as in corals

metamorphic rock the product of the interaction of metamorphic processes on a parent rock

metamorphism effect of agencies such as heat, pressure and chemically active fluids, including magmas, on rocks present within the Earth's crust

microstructure detailed form of a skeleton which is best studied by the use of a microscope

morphology scientific study of the form and structure of an organism

mould an impression of the original skeleton or soft parts

mudstone a sedimentary rock composed of fine-grained materials, similar to shale but more massive

ossicle a single calcareous plate of an echinoderm

outcrop exposed area of a particular rock type

radioactive decay the breakdown of certain kinds of isotopes at a constant rate

schizodont dental structure of certain bivalves (clams) in which two large, divergent teeth occur on right valve

sediment deposit formed from particles derived from the erosion of rocks or by the accumulation of organic materials

sedimentary rock consolidated sediment

septate central cup area of skeleton being divided by radial partitions (coral)

septum vertical division of corallite (corals) or wall between spaced planes

silicified sediment a sediment with a high silica content, the silica taking the place of earlier materials

spicular made up of numerous minute structures, *see* spicules

spicules skeletal elements having one or more axes, composed of calcite or silica, as in sponges

spinose covered in spines or elongate projections

spiralium spiral support structure found in brachiopods

stipe the single branch of a graptolite colony

stratigraphy the description, correlation and classification of bedded rocks, such as sediments and some volcanics

substrate rock surface or upper layers of sediment on which organisms live

suture line of contact between septum and shell wall (ammonites)

symbiotic association the mutually beneficial relationship of two species

taxodont dental structure of certain bivalves (clams) where numerous comb-like teeth occur along hinge line

test the hard, calcareous wall of certain invertebrates such as sea urchins

thecal wall epitheca, outer wall of coral skeleton

Books to read

British Palaeozoic Fossils (1969), British Mesozoic Fossils (1975), British Caenozoic Fossils (1975).
  British Museum (Natural History) London.
Face of the Earth, Dury, G. H., Penguin, 1970.
Geological Time, Kirkaldy, J. F., Oliver and Boyd, 1971.
The Hamlyn Guide to Minerals, Rocks and Fossils, Hamilton, W. R., Woolley, A. R. and Bishop, A. C.,
  Hamlyn, 1974.
Invertebrate Fossils, Moore, R. C., Lalicker, C. G. and Fischer, A. G., McGraw Hill, New York, 1952.
Prehistoric Animals, Cox, C. B., Hamlyn, 1969.
Principles of Palaeoecology, Ager, D. V., McGraw Hill, 1963.
Treatise on Invertebrate Palaeontology, Moore, R. C. (ed.), Geological Society of America and University of
  Kansas Press, 1954—

Acknowledgements

Photographs
Heather Angel, Hindhead 11 bottom, 42 top right, 48, 64 bottom, 69 top, 71 bottom, 77, 79
bottom, 85 top, 99; Ardea — Peter Green 17, 18 left, 18 right, 106 bottom, 107; Ardea — Imitor
39 top, 43 top, 71 top; Ardea — P. Morris 65, 88 top, 105, 117; D. Bayliss, Kingston–upon–
Thames title page, 12 left, 13 top, 23 bottom, 26, 27, 28 bottom, 29, 30, 31, 32 bottom, 34
top, 35 left, 38, 39 bottom, 42 top left, 44, 50 bottom, 52 top, 52 left, 54, 58, 59, 61, 62
bottom, 63 top, 66 bottom, 67, 72, 73 left, 75 top, 75 bottom, 76, 78, 79 top, 80, 81 top, 81
bottom, 83, 86, 87 top, 91 bottom, 93 top, 93 bottom, 94 top, 94 bottom, 95 top, 95 bottom,
97 top, 97 bottom, 98 top, 98 bottom, 100, 101 top, 101 bottom, 106 top, 109, 115, 118;
Bruce Coleman — Jane Burton endpapers, 104 bottom; Bruce Coleman — Peter Fisher 69
bottom; Bruce Coleman — Charlie Ott 12 right; Bruce Coleman — Oxford Scientific Films 43
bottom; Bruce Coleman — Leonard Lee Rue 13 bottom; Trevor D. Ford, Leicester 23 top;
Hunterian Museum, University of Glasgow 50 top, 60, 63 bottom, 64 top, 103 top; Imitor,
Bromley 14, 19, 41, 62 top; W. J. Kennedy, Oxford 32 top, 33, 34 bottom, 35 right, 37 top,
37 bottom, 52 bottom, 57, 66 top, 85 bottom, 114; T. Kooreman, Apeldoorn 70; R. T. J.
Moody, Kingston–upon–Thames 11 top, 11 right, 18 top, 20, 24, 28 top, 73 right, 82, 84, 88
bottom, 104 top, 111 top, 111 bottom, 112, 116 left; M. D. Muir, London 22 top; Oxford
Scientific Films, Long Hanborough 42 bottom; R. B. Rickards, Cambridge 40, 87 bottom, 89;
E. P. F. Rose, London 96 left, 96 right, 102; Shell, London 116 right; David J. Siveter,
Leicester 44 left, 44 right, 46; P. Wellenhoffer, Germany 91 top; R. C. L. Wilson, Berkhamp-
stead 22 bottom, 90, 108, 119.

The photographs credited to the Hunterian Museum, University of Glasgow are reproduced
by permission of the Museum and are all of items in their collection.

Index

Numbers in italic refer to illustrations